Electron Beam Processing of Materials

Electron Beam Processing of Materials

Editor

Katia Vutova

MDPI • Basel • Beijing • Wuhan • Barcelona • Belgrade • Manchester • Tokyo • Cluj • Tianjin

Editor
Katia Vutova
Bulgarian Academy of Sciences
Bulgaria

Editorial Office
MDPI
St. Alban-Anlage 66
4052 Basel, Switzerland

This is a reprint of articles from the Special Issue published online in the open access journal *Materials* (ISSN 1996-1944) (available at: https://www.mdpi.com/journal/materials/special_issues/Electron_Beam_Processing_Materials).

For citation purposes, cite each article independently as indicated on the article page online and as indicated below:

LastName, A.A.; LastName, B.B.; LastName, C.C. Article Title. *Journal Name* **Year**, *Volume Number*, Page Range.

ISBN 978-3-0365-4933-0 (Hbk)
ISBN 978-3-0365-4934-7 (PDF)

© 2022 by the authors. Articles in this book are Open Access and distributed under the Creative Commons Attribution (CC BY) license, which allows users to download, copy and build upon published articles, as long as the author and publisher are properly credited, which ensures maximum dissemination and a wider impact of our publications.

The book as a whole is distributed by MDPI under the terms and conditions of the Creative Commons license CC BY-NC-ND.

Contents

About the Editor vii

Preface to "Electron Beam Processing of Materials" ix

Tatyana Olshanskaya, Vladimir Belenkiy, Elena Fedoseeva, Elena Koleva and Dmitriy Trushnikov
Application of Dynamic Beam Positioning for Creating Specified Structures and Properties of Welded Joints in Electron-Beam Welding
Reprinted from: *Materials* **2020**, *13*, 2233, doi:10.3390/ma13102233 1

Florian Pixner, Fernando Warchomicka, Patrick Peter, Axel Steuwer, Magnus Hörnqvist Colliander, Robert Pederson and Norbert Enzinger
Wire-Based Additive Manufacturing of Ti-6Al-4V Using Electron Beam Technique
Reprinted from: *Materials* **2020**, *13*, 3310, doi:10.3390/ma13153310 17

Thomas Schnabel, Marius Cătălin Barbu, Eugenia Mariana Tudor and Alexander Petutschnigg
Changing in Larch Sapwood Extractives Due to Distinct Ionizing Radiation Sources
Reprinted from: *Materials* **2021**, *14*, 1613, doi:10.3390/ma14071613 41

Jean-Pierre Bellot, Julien Jourdan, Jean-Sébastien Kroll-Rabotin, Thibault Quatravaux and Alain Jardy
Thermal Behavior of Ti-64 Primary Material in Electron Beam Melting Process
Reprinted from: *Materials* **2021**, *14*, 2853, doi:10.3390/ma14112853 49

Hanlin Peng, Weiping Fang, Chunlin Dong, Yaoyong Yi, Xing Wei, Bingbing Luo and Siming Huang
Nano-Mechanical Properties and Creep Behavior of Ti6Al4V Fabricated by Powder Bed Fusion Electron BeamAdditive Manufacturing
Reprinted from: *Materials* **2021**, *14*, 3004, doi:10.3390/ma14113004 63

Mirela Brașoveanu and Monica R. Nemțanu
Temperature Profile in Starch during Irradiation. Indirect Effects in Starch by Radiation-Induced Heating
Reprinted from: *Materials* **2021**, *14*, 3061, doi:10.3390/ma14113061 77

Katia Vutova, Vladislava Stefanova, Vania Vassileva and Milen Kadiyski
Behaviour of Impurities during Electron Beam Melting of Copper Technogenic Material
Reprinted from: *Materials* **2022**, *15*, 936, doi:10.3390/ma15030936 87

Arsen Muslimov and Vladimir Kanevsky
Cathodoluminescent Analysis of Sapphire Surface Etching Processes in a Medium-Energy Electron Beam
Reprinted from: *Materials* **2022**, *15*, 1332, doi:10.3390/ma15041332 99

Matthias Moschinger, Florian Mittermayr and Norbert Enzinger
Influence of Beam Figure on Porosity of Electron Beam Welded Thin-Walled Aluminum Plates
Reprinted from: *Materials* **2022**, *15*, 3519, doi:10.3390/ma15103519 113

Alexander A. Svintsov, Maxim A. Knyazev and Sergey I. Zaitsev
Calculation of the Absorbed Electron Energy 3D Distribution by the Monte Carlo Method, Presentation of the Proximity Function by Three Parameters α, β, η and Comparison with the Experiment
Reprinted from: *Materials* **2022**, *15*, 3888, doi:10.3390/ma15113888 127

Katia Vutova, Vladislava Stefanova, Vania Vassileva and Stela Atanasova-Vladimirova
Recycling of Technogenic CoCrMo Alloy by Electron Beam Melting
Reprinted from: *Materials* **2022**, *15*, 4168, doi:10.3390/ma15124168 . . . . **139**

About the Editor

Katia Vutova

Institute of Electronics, Bulgarian Academy of Sciences, Sofia, Bulgaria

Interests: electron and ion beam technologies; nanotechnology; mathematical modeling and application of mathematics in physics; interaction of electron and ion beams with materials; technology optimization.

Preface to "Electron Beam Processing of Materials"

The Special Issue on "Electron Beam Processing of Materials" brings together scientists working at universities, research institutes, laboratories, and various industries to discuss state-of-the-art research on material processing using electron beam methods for different applications. The methods and devices based on the use of electron beams for material processing are high-tech, environmentally friendly, resource-preserving technologies and devices that are key for developing high-quality competitive products. They are the foundation of technical progress in micro- and nano-electronics, in the production of novel materials, in the creation of new designs of instruments and precision machinery, and in the development of technologies and systems based on electron beam.

This Special Issue is a timely approach to survey recent progress in the development and optimization of electron beam applications in modern, ecological, conventional, and nonconventional methods for material processing. The articles presented in this Special Issue cover various topics, including metal melting and welding, additive manufacturing, electron beam irradiation, electron beam lithography, process modeling, etc.

Katia Vutova
Editor

Article

Application of Dynamic Beam Positioning for Creating Specified Structures and Properties of Welded Joints in Electron-Beam Welding

Tatyana Olshanskaya [1], Vladimir Belenkiy [1], Elena Fedoseeva [1], Elena Koleva [2] and Dmitriy Trushnikov [1,*]

1. Department of Welding Production, Metrology and Material Technology, Perm National Research Polytechnic University, Perm 614990, Russia; tvo66@mail.ru (T.O.); vladimirbelenkij@yandex.ru (V.B.); elena.fedoseeva.79@mail.ru (E.F.)
2. Faculty of Physics, University of Chemical Technology and Metallurgy, 1756 Sofia, Bulgaria; eligeorg@abv.bg
* Correspondence: trdimitr@yandex.ru; Tel.: +7-9-194785031

Received: 15 April 2020; Accepted: 8 May 2020; Published: 13 May 2020

Abstract: The application of electron beam sweep makes it possible to carry out multifocal and multi-beam welding, as well as combine the welding process with local heating or subsequent heat treatment, which is important when preparing products from thermally-hardened materials. This paper presents a method of electron beam welding (EBW) with dynamic beam positioning and its experimental-calculation results regarding the formation of structures and properties of heat-resistant steel welded joints (grade of steel 20Cr3MoWV). The application of electron beam oscillations in welding makes it possible to change the shape and dimensions of welding pool. It also affects the crystallization and formation of a primary structure. It has been established that EBW with dynamic beam positioning increases the weld metal residence time and the thermal effect zone above the critical A_3 point, increases cooling time and considerably reduces instantaneous cooling rates as compared to welding without beam sweep. Also, the difference between cooling rates in the depth of a welded joint considerably reduces the degree of structural non-uniformity. A bainitic–martensitic structure is formed in the weld metal and the thermal effect zone throughout the whole depth of fusion. As a result of this structure, the level of mechanical properties of a welded joint produced from EBW with dynamic electron beam positioning approaches that of parent metal to a greater extent than in the case of welding by a static beam. As a consequence, welding of heat-resistant steels reduces the degree of non-uniformity of mechanical properties in the depth of welded joints, as well as decreases the level of hardening of a welded joint in relation to parent metal.

Keywords: electron-beam welding; welded metal structure; dynamic positioning of an electron beam; electron beam

---

1. Introduction

Electron beam welding (EBW) is being increasingly implemented in various applications [1], including the manufacturing of essential products. In some cases, EBW is used at the final stage of manufacturing the products from thermally-hardened materials with a given set of properties, which makes it difficult or impossible to carry out subsequent heat treatment of welded joints. At the same time, welded joint properties must be similar to those of the parent product material [2]. It is important to note that EBW can have the desired effect not only on solid and high-melting alloys but also on thermally hardened materials. High-quality joints are possible from the local effects of temperature [3,4]. The level of hardening of welded joints from heat-resistant steels in products for which there is only a low tempering rate is high enough. It is commonly known that the physical weldability of heat-resistant steels is hampered by the tendency of welded joints to develop cold cracks and soften metals in the thermal effect zone (TEZ). Therefore, products should be locally or preliminarily heated as part of the welding process. This decreases the temperature difference in the welding zone and peripheral areas, reducing metal stresses. The metal cooling rate decreases, and after welding, an increasingly large amount of austenite transforms to martensite at high temperature when the metal is flexible. Stresses occurring due to the difference in the volumes of these phases will decrease, and cold cracks will be less probable. It should be taken into account during heating that abnormally high temperatures result in the formation of coarse ferrite–perlite structures that do not ensure required long-term strength and impact toughness of welded joints. It is possible to reduce the risk of cold cracking by tempering parts at the temperature of 150–200 °C immediately after welding for several hours. During this time, the transformation of the residual austenite to martensite will be completed, and most of the hydrogen dissolved in it will be removed from the metal.

The softening of heat-resistant steels in TEZ also depends on the temperature-time parameters of welding. An increase in heat input enlarges a soft interlayer in TEZ, which may disrupt rigid weld joints in the course of operation, especially under bending loads. Also, structural phase transformations taking place in the welding zone mainly depend on the temperature–time parameters: degree of heating, heat distribution, heating, and cooling rates [5–9].

Due to the high concentration of energy in the electron beam affected zone, EBW is characterized by significant heating and cooling rates, as well as by high temperature gradient values. This causes a considerable non-uniformity of the temperature field that, in turn, leads to chemical, structural, and mechanical non-uniformity of a welded joint [3,10,11].

It is possible to regulate thermal cycles in EBW in order to produce a specified structure of a welded joint though controlling the thermal power of an electron beam due to its dynamic positioning [5]. The application of beam sweep makes it possible to carry out multifocal and multi-beam welding when welding is performed simultaneously in several treated areas [6,7], as well as combine the welding process with local heating or subsequent heat treatment.

When developing an EBW technology for structures from thermally hardened materials, there is the problem of choosing energy parameters of a welding mode. An additional problem is choosing the type and parameters of EBW to reproduce a specified thermal cycle, ensuring the formation of structures and properties of welded joints [10] close to those of parent material.

By using the example of aluminum alloy, we have established the main regularities of the effect of three-bath EBW parameters (beam current, welding speed, relative pulse duration, distance between points) on geometric characteristics of welds (weld depth, width, fusion shape, and fullness coefficients) and obtained regression dependencies. The criteria characterizing the formation of defect-free welds have been determined. The method for determining optimal modes of three-bath EBW for aluminum alloys by drawing nomograms and solving a system of equations from regression dependencies [5,7,12] has been proposed. Also, the effect of dynamic deflection of an electron beam on the structure and properties of welded joints from heterogeneous materials has been studied using the example of steel-bronze [9,13,14]. Based on the studies, we have developed technological recommendations to reduce structural and mechanical non-uniformity of welded joints from steel and bronze.

To the best of our knowledge, there are no studies on the effect of dynamic beam deflection in welding on the structure and properties of welded joints from heat-hardenable medium-alloy steels. However, EBW with dynamic beam positioning aimed at producing equal-strength heat-resistant steel welded joints and improving their quality continues to be of particular relevance and importance.

The purpose of this work was to study and develop a technology aimed at reducing the level of hardening of welded joints of heat-resistant steels in products for which only low-temperature tempering was provided. We present the results of studies on the effect of EBW with dynamic beam positioning on the formation of structures and properties of heat-resistant steel welded joints. The studies were conducted to develop a technology aimed at reducing the level of hardening of heat-resistant steel welded joints in products for which only low-temperature tempering is provided.

The EBW process with a sweep along the trajectory ensuring concurrent heating of welded edges was studied. The application of electron beam oscillations in welding makes it possible to change welding pool shape and dimensions, as well as affect the crystallization and formation of a primary structure. The use of dynamic positioning of an electron beam with heating of welded joint edges in EBW increases the width of a fusion zone, which affects the crystallization process and decreases cooling rates and their range of change in depth. This reduces the degree of non-uniformity of macro- and microstructure in welded joints.

2. Materials and Methods

This work includes experimental and calculation studies on the effect of dynamic positioning of an electron beam in EBW on the structure and properties of welded joints from the 20Cr3MoWV heat-resistant steel. The chemical composition of this grade of steel is given in Table 1.

Table 1. Chemical composition of the 20Cr3MoWV steel, wt %.

Fe	C	Mn	Si	Cr	Mo	V	W	Ni	P	S	Cu
								Not More than			
Base	0.15–0.23	0.25–0.5	0.17–0.37	2.8–3.3	0.35–0.55	0.6–0.85	0.6–0.85	0.03	0.03	0.025	0.025

A comparative analysis of the resulting structure and properties of a welded joint was conducted in EBW without beam oscillations, with X-shaped oscillations and with combined beam oscillations along the trajectory simulating multi-beam welding. When forming this trajectory simulating multi-beam welding, an electron beam on the metal surface was dynamically positioned such that it affected three zones, thereby forming two parallel lines on both sides of the joint at some distance from it and a point located on the joint line. The trajectory of combined electron beam oscillation is presented in Figure 1.

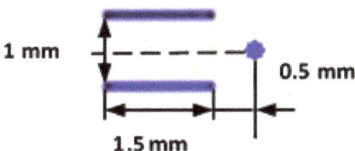

Figure 1. Trajectories of dynamic electron beam positioning.

For the selection of an electron beam positioning trajectory and its parameters, it was considered that the number of points along which the electron beam moves increases, while the depth of fusion decreases. In this case, the purpose was to obtain a fusion depth of up to 10 mm. To increase the length of a welding pool, beam oscillations along the joint were selected, and to increase its width, longitudinal oscillations were shifted in relation to the joint. The third electron beam exposure zone, a point located on the joint line, ensured the predetermined depth of fusion. The distance between the electron beam exposure zones was to have one general fusion channel formed by the electron beam in the metal (the size of the oscillation path must be small enough so that the area of the beam impact is limited to one common penetration channel). The parameters of this trajectory of dynamic positioning of an electron beam were determined based on preliminary calculations using thermal models [5,6,9]. The following parameters were determined: electron beam power—4000 W, welding speed—5 mm/s, electron beam operation time in each zone—250 µs, and longitudinal oscillation frequency—1000 Hz. The sharp focusing mode was used. The electron beam diameter (the size of an area where 95% of the electron beam energy is concentrated) was measured using the procedure and equipment described previously [15,16] and amounted to 0.35 mm (beam power—2000 W) and 0.5 mm (beam power—4000 W) on the surface of the workpiece.

There are intense convective flows in the weld pool. At the same time, the structure of the material is determined by thermal processes and depends on thermal cycles. In this work, only the thermal problem is solved, and the convective flows are not taken into account.

To implement the described trajectory, the electron gun of the ELA-6VCH power unit by SELMI (Ukraine) was upgraded through installing a high-speed deflecting system, where a signal on coils was sent from the outputs of a two-channel broadband amplifier. The amplifier was connected via a digital-to-analogue interface to a computer with installed software, allowing dynamic positioning of an electron beam along various types of user-defined trajectories. Heat-resistant steel 20Cr3MoWV was chosen as a weld material (Figure 2).

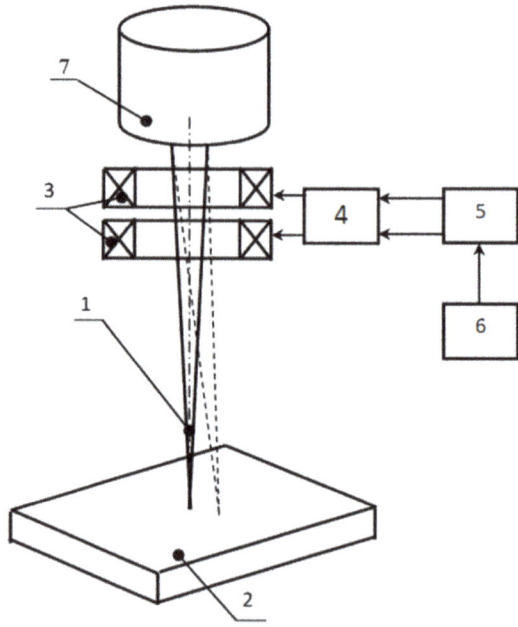

Figure 2. Scheme of the high-speed deflection system connection: 1—electron beam; 2—workpiece; 3—deflection coils; 4—broadband amplifier; 5—digital-to-analog interface; 6—computer; 7—electron beam gun.

Table 2 shows the EBW parameters for 20Cr3MoWV steel when carrying out the experimental studies.

Table 2. EBW parameters for the 20Cr3MoWV steel.

Type of EBW	Beam Power, W	Welding Speed Vweld, mm/s	Sweep Parameters	
			Frequency, Hz	Amplitude, mm
Without beam sweep	2000	5	-	-
With X-shaped oscillations	4000	5	800	2
With a sweep along the trajectory presented in Figure 1	4000	5	1000	-

The calculation studies were conducted based on mathematical models allowing the assessment of the effect of the beam oscillation trajectory and its parameters on changes in a fusion form, thermal welding cycles in the weld metal and the TEZ, instantaneous cooling rates, as well as on the conditions of formation of macro- and microstructure of welded joints. The Mathcad application program package was used for calculations.

A thermal model based on differential equations for thermal conductivity in the mobile coordinate system with a fixed source was used to construct a fusion form and thermal welding cycles. This model was obtained by an analytical method using Green's functions [1,3–5]. A standard integral solution to a thermal conductivity problem in the mobile coordinate system for an endless plate, with different types of dynamic electron beam positioning taken into account, is

$$T(x,y,z,t) = \int_{-\infty}^{\infty} \int_{-\infty}^{\infty} \int_{0}^{S} \int_{0}^{\infty} \frac{1}{8(\sqrt{\pi a(t-\tau)})^3} \exp\left(-\frac{(x-x'+V(t-\tau))^2}{4a(t-\tau)}\right) \exp\left(-\frac{(y-y')^2}{4a(t-\tau)}\right) \cdot \sum_{n=-\infty}^{\infty} \left(\exp\left(-\frac{(z-z'+2nS)^2}{4a(t-\tau)}\right) + \exp\left(-\frac{(z+z'+2nS)^2}{4a(t-\tau)}\right) \right) \cdot F(x',y',z',t) \partial x' \partial y' \partial z' \partial \tau \quad (1)$$

$$F(x',y',z',t) = \frac{\eta q}{c\rho} \cdot \delta(x-x') \cdot \delta(y-y') \cdot \delta(z-z') \cdot \delta(\tau) \quad (2)$$

where V—welding speed; S—plate thickness; $F(x',y',z',t)$—heat source function described using the Dirac delta function; x', y', z'—heat source coordinates; τ—source operation time; q—electron beam power; η—efficiency factor; c—specific heat capacity; and ρ—metal density. Heat source forms and their mathematical expressions $F(x',y',z',t)$ for different types of electron beam oscillations are presented in previous works [5,14,15].

Typically, a cooling time within the temperature range of 800–200 °C ($t_{8/2}$) and a cooling rate within the range of 600–500 °C ($w_{5/6}$) determined by a thermal welding cycle were used to analyze the microstructure of a welded joint. However, heating and cooling rates in the course of welding vary with time nonlinearly, so it is proposed to use instantaneous rates. Equations for determining instantaneous heating and cooling rates in the course of welding with different types of dynamic electron beam positioning were derived from the equation for solving thermal problems (1)

$$W(x,y,z,\tau) = \frac{dT(x,y,z,\tau)}{dt} \quad \text{при} \quad dt = \frac{dx}{V} \quad W(x,y,z,\tau) = \frac{dT(x,y,z,\tau)}{dx} V. \quad (3)$$

A mathematical model developed based on the analytical approach presented in the works [14] was used to analyze the formation of a primary macrostructure of weld metal. To ensure the mathematical setting of a model problem, the following assumptions were made: (1) the crystallization front shape represents a surface described by the crystallization isotherm equation without taking into account the sizes of a two-phase liquid–solid zone; (2) crystallites grow in the direction of temperature gradient and, consequently, their growth axes represent orthogonal trajectories to the crystallization front. The model consists of a number of equations: crystallization front, crystallite growth axis trajectories, direction angles of crystallite inclination towards coordinate planes, crystallite growth rates, as well as equations of crystallization scheme criteria and crystallization rate. Taking into account the specific form of fusion penetration in EBW, the crystallization front equation represents two systems of equations (separately for the upper and lower parts of the weld)

$$\begin{cases} \left(\frac{z}{H_1}\right)^{\omega 1} + \left(\frac{y}{P_1}\right)^{\vartheta 1} = 1 \\ \left(\frac{y}{P_1}\right)^{\eta 1} + \left(\frac{x}{L_1}\right)^{\nu 1} = 1 \\ \left(\frac{z}{H_1}\right)^{\tau 1} + \left(\frac{x}{L_1}\right)^{\mu 1} = 1 \end{cases} \begin{cases} \left(\frac{z}{H_2}\right)^{\omega 2} + \left(\frac{y}{P_2}\right)^{\vartheta 2} = 1 & \text{YOZ plane} \\ \left(\frac{y}{P_2}\right)^{\eta 2} + \left(\frac{x}{L_2}\right)^{\nu 2} = 1 & \text{XOY plane} \\ \left(\frac{z}{H_2}\right)^{\tau 2} + \left(\frac{x}{L_2}\right)^{\mu 2} = 1 & \text{XOZ plane} \end{cases} \quad (4)$$

Each system equation describes the isotherm of crystallisation for a relevant coordinate plane, Figure 3.

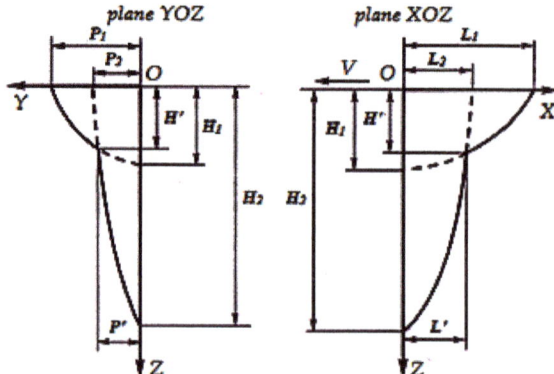

Figure 3. Schematic division of the crystallization front into components: P', H', and L'—intersection point coordinates for the two curves (inflection point); P_2, H_1, and L_2—values at which the curves intersect relevant coordinate axes; P_1—corresponds to the value of the weld half-width, H_2—corresponds to the weld depth; L_1—numerically equal to the length of the welding bath on the surface.

The coefficients and parameters of the system of crystallization front Equation (3) were determined by approximating the numerical values (x_i, y_i, z_i) of the isothermal crystallization surface obtained in solving the EBW thermal problem (1), with this system of equations. At the same time, coefficients ω, θ, η, ν, τ, and μ may have any non-integer value greater than 1. Equations for calculating the shape of crystallite axes make it possible to construct crystallite axis projections onto coordinate planes and assess a primary macrostructure of weld metal from a qualitative point of view. The values of crystallite growth origin coordinates (x_0, y_0, z_0) are determined to take into account the system Equation (3)

$$\begin{cases} y = \left[\frac{H^\omega \cdot (\theta^2 - 2\theta)}{P^\theta \cdot (\omega^2 - 2\omega)}\left(z^{2-\omega} - z_0^{2-\omega}\right) + y_0^{2-\theta}\right]^{1/(2-\theta)} & YOZ \text{ plane} \\ x = \left[\frac{P^\eta \cdot (\nu^2 - 2\nu)}{L^\nu \cdot (\eta^2 - 2\eta)}\left(y^{2-\eta} - y_0^{2-\eta}\right) + x_0^{2-\nu}\right]^{1/(2-\nu)} & XOY \text{ plane} \\ x = \left[\frac{H^\tau \cdot (\mu^2 - 2\mu)}{L^\mu \cdot (\tau^2 - 2\tau)}\left(z^{2-\tau} - z_0^{2-\tau}\right) + x_0^{2-\mu}\right]^{1/(2-\mu)} & XOZ \text{ plane} \end{cases} \quad (5)$$

Changes in the inclination angles of crystallite axes characterize the spatial orientation of crystallites and allow the numerical determination of changes in their growth rate (Figure 4). To analyze the primary macrostructure, the most indicative are the changes in the inclination angles of crystallite axes growing at different depths to the weld axis in the horizontal plane (α) and the vertical plane (γ).

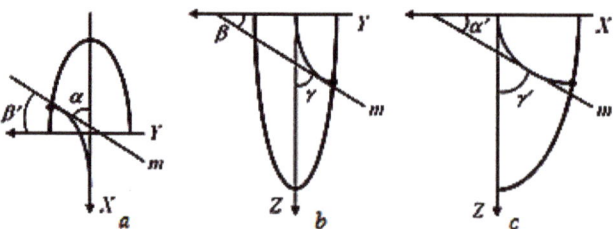

Figure 4. Direction angles of the tangents (m) to the crystallite axis: (**a**) α and β' in the XOY plane, (**b**) β and γ in the YOZ plane, (**c**) α' and γ' in the XOZ plane.

Changes in the growth rate of crystallites along the weld width in the horizontal ($V\alpha$) and vertical ($V\gamma$) planes were determined as

$$V\alpha = V\left[1 + \frac{x_0^2 \cdot \eta^2}{y_0^2 \cdot v^2} Ky^{2\eta-2}(1-Ky^\eta)^{\frac{2}{v}-2}\right]^{-\frac{1}{2}}, \qquad (6)$$

$$V\gamma = V\left[1 + \frac{H^{2\omega} \cdot \theta^2}{P^{2\theta} \cdot \omega^2} \frac{(Ky \cdot y_0)^{2\theta-2}}{z_0^{2\omega-2}}(1-Ky^\theta)^{\frac{2}{\omega}-2}\right]^{-\frac{1}{2}}, \qquad (7)$$

where V—welding speed, $Ky = y/y_0$—dimensionless coordinate.

The integral criteria of the $K\alpha$ and $K\gamma$ crystallization scheme make it possible to assess resulting macrostructures in EBW. The $K\alpha$ criterion characterizes the preferred direction of crystallite axes along the weld width in the horizontal plane, and the $K\gamma$ criterion characterizes the preferred direction of crystallite axes along the weld width in the vertical plane

$$K\alpha = \int_0^1 \text{arctg}\left[\frac{x_0 \cdot \eta}{y_0 \cdot v} Ky^{\eta-1}(1-Ky^\eta)^{-\frac{v-1}{v}}\right] dKy, \qquad (8)$$

$$K\gamma = \int_0^1 \text{arctg}\left[\frac{H^\omega \cdot \theta}{P^\theta \cdot \omega} \frac{(Ky \cdot y_0)^{\theta-1}}{z_0^{\omega-1}}(1-Ky^\theta)^{-\frac{\omega-1}{\omega}}\right] dKy. \qquad (9)$$

A crystallite growth rate is quantitatively assessed using the two integral criteria of crystallization rate—$KV\alpha$ and $KV\gamma$, characterizing the total value of relative crystallite growth rate along the weld width in the horizontal and vertical planes

$$KV\alpha = \int_0^1 \left[1 + \frac{H^{2\omega} \cdot \theta^2}{P^{2\theta} \cdot \omega^2} \frac{(Ky \cdot y_0)^{2\theta-2}}{z_0^{2\omega-2}}(1-Ky^\theta)^{\frac{2}{\omega}-2}\right]^{-1/2} dKy, \qquad (10)$$

$$KV\gamma = \int_0^1 \left[1 + \frac{H^{2\omega} \cdot \theta^2}{P^{2\theta} \cdot \omega^2} \frac{(Ky \cdot y_0)^{2\theta-2}}{z_0^{2\omega-2}}(1-Ky^\theta)^{\frac{2}{\omega}-2}\right]^{-1/2} dKy. \qquad (11)$$

The application of integral criteria of crystallization scheme and crystallization rate makes it possible to construct diagrams characterizing the macrostructure of weld metal.

To analyze the emerging microstructure of welded joints, a method was used based on the construction of a series of structural diagrams depending on the cooling rate. When plotting structural diagrams, the regression models (obtained using artificial neural networks) of the transformation of supercooled austenite under continuous cooling were used [17]. These regression equations determine the type of microstructure formed after cooling with four dichotomous variables containing information on the presence of ferrite, perlite, bainite, and martensite in the structure

$$X(\%) = \begin{cases} 0 \text{ при } W_X = 0 \\ 0 \text{ при } U_X \leq 0 \\ U_X \text{ при } U_X > 0 \end{cases} \qquad (12)$$

$$U_X = a_0 + a_1 C + a_2 Mn + a_3 Si + a_4 Cr + a_5 Ni + a_6 Mo + a_7 V + a_8 T_A + a_8 W_{\text{о х л}}$$
$$+ a_{10} CV_r^{0.25} + a_{11} W_f + a_{12} W_p + a_{13} W_b + a_{14} W_m$$

where C, Mn, Si, Cr, Ni, Mo, V—weight fractions of alloying elements; $a_0, a_1, \ldots, a_{14}$—coefficients obtained by regression analysis; U_X—volume fraction of the structural component; X—type of the structural component; T_A—austenitizing temperature, °C; W_{cool}—cooling rate, °C/min; W_X—dichotomous variables.

Dichotomous variables are designed to determine the probability of specific microstructural components at a given constant cooling rate and austenitizing temperature

$$\begin{aligned} W_X &= \exp(S_X)/(1+\exp(S_X)), \\ S_X = b_{0X} + b_{1X}C &\quad +b_{2X}Mn + b_{3X}Si + b_{4X}Cr + b_{5X}Ni + b_{6X}Mo + b_{7X}V + b_{8X}Cu \\ &\quad +b_{9X}T_A + b_{10X}W_{okh.\pi} \end{aligned} \quad (13)$$

where $b_{0X}, b_{1X}, \ldots b_{10X}$—coefficients obtained by regression analysis.

To determine the quantitative composition of the microstructure formed in the course of welding, several diagrams were constructed based on these equations, separately for each section of a welded joint. When constructing diagrams, austenitizing temperature and cooling rates are pre-set individually. The austenitizing temperature is equal to the maximum heating temperature of this area as far as the TEZ different parts are concerned. As for the weld, the T_A is 1350 °C. The range of cooling rates was chosen based on maximum possible design instantaneous cooling rates according to the formulas obtained from the expression (3). Diagrams were constructed in the coordinates "% of structural components (X)—cooling rate (W_{cool})". A criterion for determining the structural composition is the maximum instantaneous cooling rate obtained from calculations for a given part of a welded joint.

The experimental studies were conducted with regard to the specimens welded according to the modes shown in Table 2 and included the metallographic analysis of the macro- and microstructure of welded joints, determination of hardness, and mechanical properties in standard static tension tests. The metallographic analysis was conducted using the Altami MET 1T optical microscope and the VideoTest Metall image analysis software system. The surface of microslices was treated alternately with two reagents and multiple repolishing (the first reagent based on nitric acid, and the second one based on picric acid). Hardness was determined using the PMT-3 instrument and the Tukon 2500 Vickers hardness tester.

3. Results and Discussion

The results of experimental and calculation studies of the weld metal crystallization process in EBW (20Cr3MoWV steel) are presented in Figure 5, which also presents the weld macrostructure (a), the shape of a crystallizing part of the welding bath with crystallite axis projections (b), and the diagram of changes in the macrostructure shape and crystallization scheme (c). The shape and sizes of the crystallizing part were determined by the thermal model (1, 2); the crystallite axis projections were determined by the equations (4, 5); the diagrams of changes in macrostructure shape and crystallization scheme were based on the calculations of integral crystallization criteria (6–11).

The results show that the similar depth of fusion welding with combined beam oscillation trajectory leads to an increase in the sizes of the crystallizing part of the welding bath, in relation to welding without beam sweep and with x-shaped oscillations. At the same time, the weld shape in cross-section is close to that of the weld produced by welding with x-shaped oscillations and differs by its width. An increase in the sizes of the crystallizing part affects the crystallization process and the macrostructure formation; the width of a central zone with equiaxed grains increases, and there is no flat growth pattern for columnar crystallites.

The calculations of thermal cycles and instantaneous cooling rates were made for the following parts of welded joints:

- In the depth of a welded joint: for the upper—0.1H, middle—0.5H and root—0.9H parts;
- In the width of a welded joint: for the weld metal, for the overheating area, temperature—1350 °C, for the full recrystallisation area, temperature—1000 °C.

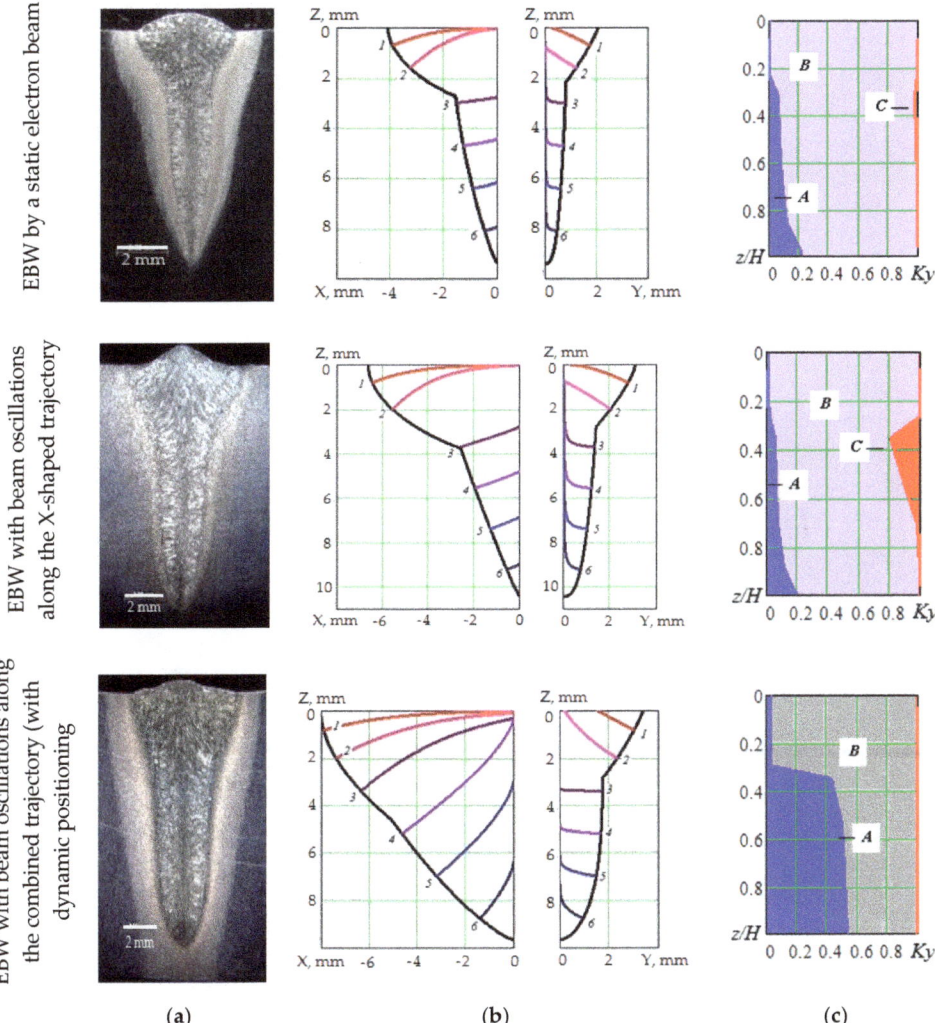

Figure 5. Weld macrostructure (**a**), shape of the crystallizing part of the welding bath with crystallite axis projections in the XOZ and YOZ coordinate planes (**b**), and diagrams of changes in macrostructure shape and crystallization pattern in provisional coordinates in the z/H depth and in the $Ky = y/y_0$ (**c**) weld width: (A—equiaxial structure, B—spatial growth of columnar crystallites, C—flat growth of columnar crystallites).

An example of calculated thermal cycles and instantaneous cooling rates for the TEZ overheating area is given in Figure 6.

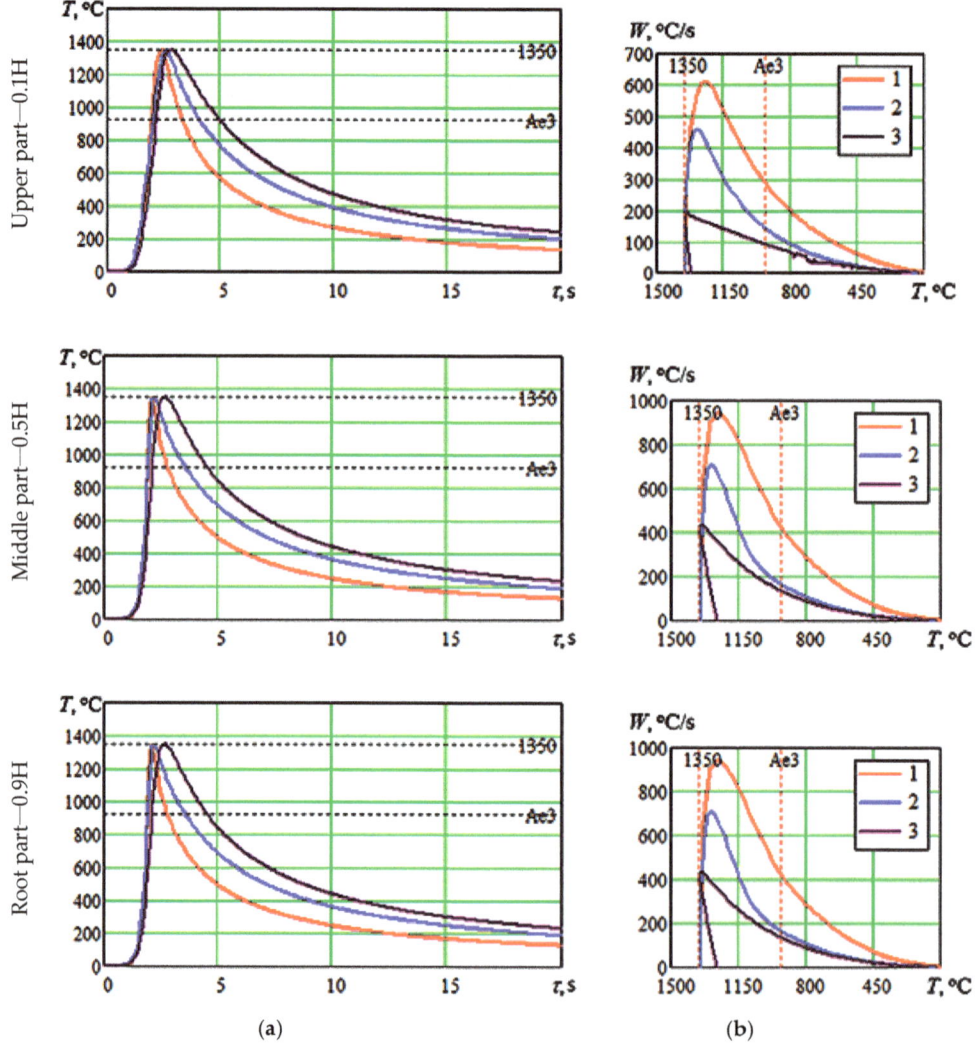

Figure 6. Calculated thermal cycles (**a**) and instantaneous cooling rates (**b**) in the TEZ overheating area (maximum heating temperature—1350 °C) for different weld depths: 1—welding without beam sweep; 2—welding with x-shaped beam oscillations; 3—welding with combined beam sweep.

The calculations show that the extension of a fusion zone in welding with beam oscillations along the combined trajectory contributes to the fact that the time of metal residence in its liquid state increases and the range of maximum instantaneous cooling rates for crystallized metal decreases. Thus, in the weld metal produced from welding without beam sweep, the maximum instantaneous cooling rate at the temperature of 1350 °C varies in the depth of fusion from 900 to 3600 °C/s; in the weld metal produced from welding with x-shaped beam oscillations, it varies from 460 to 2700 °C/s, and in the weld metal produced from welding with combined beam sweep—from 220 to 1250 °C/s. In the thermal effect zone, the time of metal residence increases above the critical A_3 point (Figure 6), and the level of maximum instantaneous cooling rates, as well as their range, decreases as compared to welding without beam oscillations and with x-shaped oscillations. In the TEZ part heated up to the maximum temperature of 1350 °C, the range of instantaneous cooling rates is 610–2200 °C/s when welding by a static beam and 190–980 °C/s when welding with combined beam sweep. Maximum instantaneous cooling rates decrease in the TEZ parts heated up to lower temperatures, but more than the A_3 point. In the area heated up to the maximum temperature of 1000 °C when welding by a static beam, their range is 200–750 °C/s, and when welding with combined beam sweep, their range is 85–480 °C/s.

A decrease in the difference between cooling rates in the depth of a welded joint creates conditions for reducing the degree of structural non-uniformity. To determine the quantitative composition of structural components in welded joints, a calculation method for analyzing the emerging microstructure of welded joints was used, and metallographic studies were conducted.

To calculate the volume fraction of formed structural components in welded joints, structural diagrams were constructed depending on the cooling rate (12–13) in the coordinates "% of structural components (X) —cooling rate (W_{cool})". When constructing the diagrams, the following austenitizing temperatures were pre-set: for the weld metal and the TEZ overheating area, the T_A = 1350 °C, for the TEZ full recrystallization area, the T_A = 1100 °C. A percentage ratio of structural components is determined according to the range of maximum instantaneous cooling rates for a given area. An example of determining the volume fraction of structural components is presented in Figure 7.

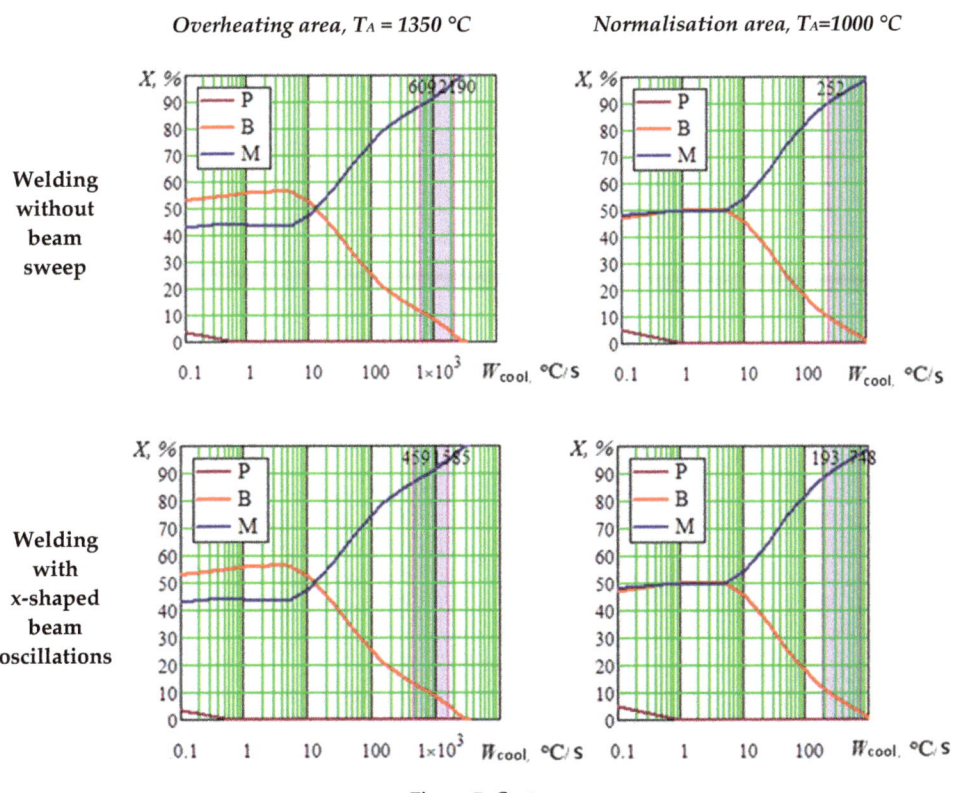

Figure 7. Cont.

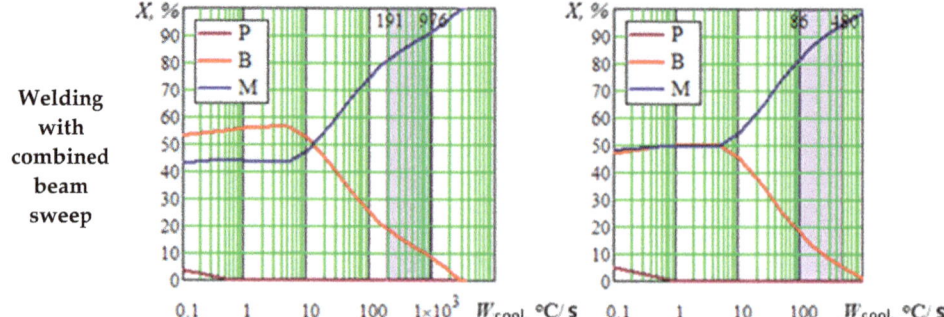

Figure 7. Structural diagrams for the TEZ two parts (a welded joint from the 20H3MVF steel): P—perlite, B—bainite, M—martensite (the diagrams show the range of instantaneous cooling rates in the depth of welded joints).

According to the presented calculation results, a practically martensitic structure should be formed in the TEZ when welding by a static beam: in the upper part, the amount of bainite does not exceed 10%, while in the lower part of the weld, it does not exceed up to 2%. The application of x-shaped beam oscillations should lead to an increase in bainite in the upper part up to 10–12%, and in the root part up to 2–5%. When welding with combined beam oscillations, about 20% of bainite can be formed in the TEZ upper part and within 5–10% in the lower part.

As far as it is known, conventional methods of metallographic analysis do not always make it possible to clearly differentiate structural components close in morphological structure, such as lower bainite and martensite. The method of multiple polishing with reagent alternation results in detecting a relief on the surface of a slice, which creates a more comprehensive idea of the morphology of resulting structures, especially under large increases. Polarized light was additionally used to differentiate structural components in microstructure analysis. Polarized light reflects the morphology of formed structures and singles out a carbide phase (present in bainite). The carbide phase in polarized light is singled out in the form of rounded light inclusions with clear outlines. Consequently, the areas where intermediate (bainite) transformation takes place will be tinted with a luminous ring of light. Martensite in polarized light looks darker; these dark inclusions reflect its morphology—a package structure.

The quantitative assessment of structural components was carried out using the VideoTest-Metal image analysis software system. This software system singles out structural components according to their brightness range and determines the volume fraction of a highlighted phase. The analysis was conducted with regard to five fields of view. Figure 8 shows a microstructure, an example of separating phases and determining their volume fraction for the middle part of weld metal. The measurement results regarding the quantitative composition of weld metal are shown in Table 3, including errors and the composition calculated according to the above procedure for comparison.

The emergence of bainite in the weld metal structure and the thermal effect zone is of great importance for the mechanical properties of heat-resistant steels. When welding these steels without further heat treatment, the most optimal combination of mechanical properties will be in welded joints with mixed martensite and bainite structure.

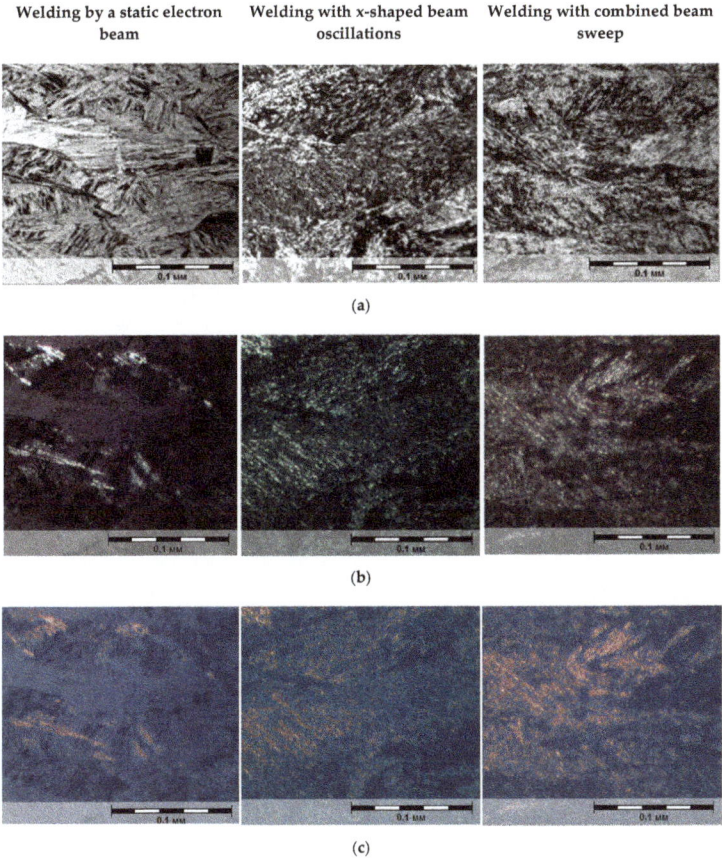

(a)

(b)

(c)

Figure 8. Weld metal microstructure in the middle (**a**) part and an example of determining a percentage ratio of structural components: polarized light (**b**), separating structural components in the VideoTest-Metal program (**c**), ×500.

Table 3. Structural composition of weld metal obtained experimentally and by calculation.

Type of Electron Beam Sweep	Distance in the Depth pf Joint	Experimental Observation	Calculation Based on Regression Dependencies
Without beam sweep	0.1H	7.3% V, 92.7% M ($\varepsilon - 5.7\%$)	9% V, 91% M
	0.5H	4.8% V, 95.2% M ($\varepsilon - 3.11\%$)	4% V, 96% M
	0.9H	2.9% V, 97.1% M ($\varepsilon - 4.35\%$)	0% V, 100% M
With X-shaped oscillations	0.1H	18.8% V, 8.2% M ($\varepsilon - 2.8\%$)	14% V, 86% M
	0.5H	14% V, 86% M ($\varepsilon - 2.5\%$)	8.5% V, 91.5% M
	0.9H	9.3% V, 90.7% M ($\varepsilon - 2.9\%$)	2% V, 98% M
Three-beam welding	0.1H	23% V, 77% M ($\varepsilon - 4\%$)	19% V, 81% M
	0.5H	20% V, 80% M ($\varepsilon - 4.2\%$)	14% V, 86% M
	0.9H	13% V, 87% M ($\varepsilon - 9\%$)	7% V, 93% M

The degree of non-uniformity in mechanical properties was assessed by the nature of variation in microhardness in the width of a welded joint for areas different in depth. Microhardness was measured in the upper, middle, and lower parts of welded joints using the Vickers method. Based on the obtained information, the average values of weld metal microhardness and root-mean-square deviation (RMSD) were determined as a measure of non-uniformity (Table 4). When welding by a static beam in the depth of a weld, there is an increase in RMSD of microhardness values, which indirectly indicates an increase in mechanical inhomogeneity. When welding with combined beam sweep and with x-shaped oscillations, the average value of weld metal microhardness decreases. RMSD values also decrease, i.e., it is arguable that the degree of mechanical inhomogeneity is lower here.

Table 4. Changes in weld metal hardness for different types of EBW.

Part	Average Hardness Values	Welding by a Static Electron Beam	Welding with x-Shaped Beam Oscillations	Welding with Combined Beam Sweep
Weld upper part	Aver. HV 0.1	481	404	418
	RMSD	94	60	66
Weld middle part	Aver. HV 0.1	479	394	413
	RMSD	127	80	67
Weld lower part	Aver. HV 0.1	488	410	394
	RMSD	136	88	64

The specimens cut out in the cross-section of a welded joint were used for tension tests. Strength and flexibility properties were determined for the weld metal. The specimens were fractured at the weakest point of a welded joint. The results are given in Table 5.

Table 5. Mechanical properties of welded joints produced with different types of EBW.

Mechanical Properties	Type of EBW			Parent Metal
	Welding by a Static Electron Beam	Welding with x-Shaped Beam Oscillations	Welding with Combined Beam Sweep	
σ_V, MPa	1535	1333	1301	822
$\sigma_{0.2}$, MPa	1397	1193	1159	75
$\sigma_{0.2}/\sigma_V$	0.91	0.894	0.882	0.75
δ_r, %	4.1	4.43	4.49	8.84

The results show that welding with x-shaped beam oscillations and with combined sweep reduces the level of strength properties. Tensile strength σ_B and constrained yield stress $\sigma_{0.2}$ have similar values. At the same time, these values are much lower than σ_T and $\sigma_{0.2}$ of the weld metal produced from welding without beam oscillations. Similar values were also obtained for relative ultimate uniform elongation δ_r. Metal flexibility margin was assessed by ultimate strength against yield strength ($\sigma_{0.2}/\sigma_V$), and if $\sigma_{0.2}/\sigma_V = 1$, metal flexibility margin was considered to be zero. As for medium-alloy steels, a ratio of yield strength against ultimate strength is permissible within the range of 0.7–0.8. Weld metal flexibility margin is slightly higher in case of welding with combined sweep.

Considering the foregoing, the proposed EBW method with dynamic beam positioning along the combined trajectory ensures a significant decrease in structural and mechanical inhomogeneity and brings the level of mechanical properties closer to that of the parent material. Consequently, this EBW method will be preferable for heat-resistant steel structures not subjected to further heat treatment.

4. Conclusions

The application of electron beam oscillations in welding makes it possible to change a welding pool shape and dimensions, as well as affect the crystallization and the formation of a primary structure. The emergence of bainite in the weld metal structure and the thermal effect zone is of great importance for the mechanical properties of heat-resistant steels. When welding these steels without further heat treatment, the most optimal combination of mechanical properties will be in welded joints with mixed martensite and bainite structure. The use of the trajectory of dynamic positioning of an electron beam with heating of edges of a welded joint in EBW leads to an increase in the width of a fusion zone.

At almost the same depth of fusion, welding with combined beam oscillation trajectory leads to an increase in the sizes of the crystallizing part of the welding bath, in relation to welding without beam sweep and with x-shaped oscillations. The weld shape in cross-section is close to that of the weld produced by welding with x-shaped oscillations and differs by its width. An increase in the sizes of the crystallizing part affects the crystallization process and the macrostructure formation: the width of a central zone with equiaxed grains increases and there is no flat growth pattern for columnar crystallites. The extension of a fusion zone increases the weld metal's residence time and the thermal effect zone above the critical A_3 point, increases cooling time, and considerably reduces instantaneous cooling rates and their range of change in depth respectively. This further reduces the degree of non-uniformity of macro- and microstructure in welded joints. As a result, welding of heat-resistant steels reduces the degree of non-uniformity of mechanical properties in the depth of welded joints, as well as reduces the level of hardening of a welded joint in relation to parent metal.

Author Contributions: Conceptualization, T.O. and E.F.; methodology, software, T.O.; validation, V.B., E.K. and E.F.; formal analysis, T.O.; investigation, T.O. and E.K.; data curation, D.T.; writing—original draft preparation, V.B.; writing—review and editing, D.T. and E.F.; visualization, D.T.; project administration, D.T.; funding acquisition, D.T. All authors have read and agreed to the published version of the manuscript.

Funding: The reported study was partially supported by the Government of Perm Krai research project no. S-26/787 from 21.12.2017, by grants from the Russian Foundation for Basic Research RFBR no. 18-08-01016 A as well as by Ministry of Education and Science of the Russian Federation at the base part of the state task grant "Development and creation of functional materials and products with specified physical, mechanical and operational properties using methods of electrophysical effects and machining" (FSNM-2020-0028).

Conflicts of Interest: The authors declare that there is no conflict of interests regarding the publication of this paper.

References

1. Koleva, E.; Mladenov, G.M.; Trushnikov, D.; Belenkiy, V. Signal emitted from plasma during electron-beam welding with deflection oscillations of the beam. *J. Mater. Process. Technol.* **2014**, *214*, 1812–1819. [CrossRef]
2. Trushnikov, D.; Koleva, E.; Mladenov, G.; Belenkiy, V.Y. Effect of beam deflection oscillations on the weld geometry. *J. Mater. Process. Technol.* **2013**, *213*, 1623–1634. [CrossRef]
3. Serajzadeh, S. Modeling of temperature history and phase transformations during cooling of steel. *J. Mater. Process. Technol.* **2004**, *146*, 311–317. [CrossRef]
4. Sun, Z.; Karppi, R. The application of electron beam welding for the joining of dissimilar metals: An overview. *J. Mater. Process. Technol.* **1996**, *59*, 257–267. [CrossRef]
5. Olshanskaya, T.V.; Fedoseeva, E.M.; Belenkiy, V.Y.; Trushnikov, D.N. Thermal model in electron beam welding with various dynamic positioning of the beam. *J. Phys. Conf. Ser.* **2018**, *1089*, 012007. [CrossRef]
6. Olshanskaya, T.V.; Salomatova, E.; Belenkiy, V.Y.; Trushnikov, D.; Permyakov, G.L. Electron beam welding of aluminum alloy AlMg6 with a dynamically positioned electron beam. *Int. J. Adv. Manuf. Technol.* **2016**, *89*, 3439–3450. [CrossRef]
7. Almonti, D.; Ucciardello, N. Design and Thermal Comparison of Random Structures Realized by Indirect Additive Manufacturing. *Materials* **2019**, *12*, 2261. [CrossRef]
8. Kaisheva, D.; Angelov, V.; Petrov, P. Simulation of heat transfer at welding with oscillating electron beam. *Can. J. Phys.* **2019**, *97*, 1140–1146. [CrossRef]
9. Olshanskaya, T.V.; Fedoseeva, E.M. Development of methods for forecasting the microstructure of the welded joint with reference to the temperature-time modes of electron-beam welding. *IOP Conf. Ser. Mater. Sci. Eng.* **2020**, *759*, 012020. [CrossRef]
10. Ding, H.; Huang, Q.; Liu, P.; Bao, Y.; Chai, G. Fracture Toughness, Breakthrough Morphology, Microstructural Analysis of the T2 Copper-45 Steel Welded Joints. *Materials* **2020**, *13*, 488. [CrossRef] [PubMed]
11. Liu, P.; Bao, J.; Bao, Y. Mechanical Properties and Fracture Behavior of an EBW T2 Copper-45 Steel Joint. *Materials* **2019**, *12*, 1714. [CrossRef] [PubMed]

12. Angella, G.; Barbieri, G.; Donnini, R.; Montanari, R.; Richetta, M.; Varone, A. Electron Beam Welding of IN792 DS: Effects of Pass Speed and PWHT on Microstructure and Hardness. *Materials* **2017**, *10*, 1033. [CrossRef] [PubMed]
13. Hsu, T.-I.; Jhong, Y.-T.; Tsai, M.-H. Effect of Gradient Energy Density on the Microstructure and Mechanical Properties of Ti6Al4V Fabricated by Selective Electron Beam Additive Manufacture. *Materials* **2020**, *13*, 1509. [CrossRef] [PubMed]
14. Prokhorov, N.N. *Physical Processes in Metals at Welding*; Metallurgy: Moscow, Russia, 1968; p. 695.
15. Elmer, J.W.; Teruya, A.T. An enhanced faraday cup for rapid determination of power density distribution in electron beams. *Weld. J.* **2001**, *80*, 288–295.
16. Palmer, T.A.; Elmer, J.W. Characterisation of electron beams at different focus settings and work distances in multiple welders using the enhanced modified Faraday cup. *Sci. Technol. Weld. Join.* **2007**, *12*, 161–174. [CrossRef]
17. Trzaska, J. Calculation of volume fractions of microstructural components in steels cooled from the austenitizing temperature. *J. Achiev. Mater. Manuf. Eng.* **2014**, *65*, 38–44.

© 2020 by the authors. Licensee MDPI, Basel, Switzerland. This article is an open access article distributed under the terms and conditions of the Creative Commons Attribution (CC BY) license (http://creativecommons.org/licenses/by/4.0/).

Article

Wire-Based Additive Manufacturing of Ti-6Al-4V Using Electron Beam Technique

Florian Pixner [1], Fernando Warchomicka [1], Patrick Peter [1,2], Axel Steuwer [3,4], Magnus Hörnqvist Colliander [5], Robert Pederson [6] and Norbert Enzinger [1,*]

1. Institute of Materials Science, Joining and Forming, Graz University of Technology, Kopernikusgasse 24, 8010 Graz, Austria; florian.pixner@tugraz.at (F.P.); fernando.warchomicka@tugraz.at (F.W.); patrick.peter@liebherr.com (P.P.)
2. Liebherr-Werk Telfs GmbH, 6410 Telfs, Austria
3. Nelson Mandela University, Port Elizabeth 6031, South Africa; axel.steuwer@um.edu.mt
4. Research Support Services, University of Malta, 2080 Msida, Malta
5. Department of Applied Physics, Chalmers University of Technology, 41296 Göteborg, Sweden; magnus.colliander@chalmers.se
6. Department of Engineering Science, University West, 46132 Trollhättan, Sweden; robert.pederson@hv.se
* Correspondence: norbert.enzinger@tugraz.at; Tel.: +43-(0)316-873-7182

Received: 6 June 2020; Accepted: 20 July 2020; Published: 24 July 2020

Abstract: Electron beam freeform fabrication is a wire feed direct energy deposition additive manufacturing process, where the vacuum condition ensures excellent shielding against the atmosphere and enables processing of highly reactive materials. In this work, this technique is applied for the α + β-titanium alloy Ti-6Al-4V to determine suitable process parameter for robust building. The correlation between dimensions and the dilution of single beads based on selected process parameters, leads to an overlapping distance in the range of 70–75% of the bead width, resulting in a multi-bead layer with a uniform height and with a linear build-up rate. Moreover, the stacking of layers with different numbers of tracks using an alternating symmetric welding sequence allows the manufacturing of simple structures like walls and blocks. Microscopy investigations reveal that the primary structure consists of epitaxial grown columnar prior β-grains, with some randomly scattered macro and micropores. The developed microstructure consists of a mixture of martensitic and finer α-lamellar structure with a moderate and uniform hardness of 334 HV, an ultimate tensile strength of 953 MPa and rather low fracture elongation of 4.5%. A subsequent stress relief heat treatment leads to a uniform hardness distribution and an extended fracture elongation of 9.5%, with a decrease of the ultimate strength to 881 MPa due to the fine α-lamellar structure produced during the heat treatment. Residual stresses measured by energy dispersive X-ray diffraction shows after deposition 200–450 MPa in tension in the longitudinal direction, while the stresses reach almost zero when the stress relief treatment is carried out.

Keywords: additive manufacturing; titanium alloys; electron beam; wire feed process; residual stresses; mechanical properties

1. Introduction

Additive manufacturing (AM) comprises different processes which deal with different materials using different heat sources to build-up structural parts. The ISO 17296-2:2015 standard provides an overview of existing AM process categories [1]. Processes can be classified according to their general principles as follows: vat photopolymerization, material jetting, binder jetting, powder bed fusion (PBF), material extrusion, directed energy deposition (DED), and sheet lamination.

For reactive materials like Ti-based alloys, the number of feasible processes is limited and the demands on the shielding environment are remarkably high to avoid atmospheric contaminations. Nowadays Ti-6Al-4V, recognized as the most popular α + β titanium alloy, is processed by various AM processes like powder-based processes, selective laser melting (SLM), electron beam melting (EBM), or wire-based DED techniques. The respective processes show certain characteristics and result in intrinsic properties of the manufactured parts. Powder-based processes are very common for AM and are leading technologies performing complex geometries and surface finishing. In AM, there is a correlation and trade-off between maximum resolution and achievable deposition rates. Due to the usage of powder in the micrometer range, the achievable deposition rates are limited, and the part size is restricted to the dimensions of working chambers. Wire-based DED processes widen the field of application and have received considerable attention due to printing of more volumetric structures with simultaneous high deposition rates. Electron beam freeform fabrication (EBF^3, EBF3, or EBFFF) is based on a wire feed DED-process using the electron beam as heat source and the suitability for processing titanium has already been demonstrated [2–5]. Since electron beam technique is based on a vacuum atmosphere (<5 × 10^{-3} mbar), it is suitable to process reactive materials like titanium. The process can be described as a near-net-shape manufacturing process with a high deposition rate (up to 2500 cm^3/h) which usually requires an additional final processing step, e.g., subtractive manufacturing [6]. The characteristic high-power density of the electron beam welding (EBW) process of up to 10^7 W/cm^2 is not required in case of AM, because there is no need of a deep penetration since dilution should be minimized [7]. This type of process is currently used for commercial purposes to produce robust structures of different titanium, tungsten, and inconel alloys, among others [8].

In the case of wire-based electron beam additive manufacturing (EBAM), there are numerous technological input parameters—e.g., (1) acceleration voltage, (2) beam current, (3) welding speed, (4) wire feed rate, (5) beam figure, and (6) focus position—which can be varied in a wide range and a proper selection of them can be challenging [9]. Another factor in process design is the way of the wire-feeding. Usually, wire arc additive manufacturing (WAAM) processes use an electric arc as energy source without electromagnetic force between the molten droplet and the substrate plate. In the case of wire-based electron beam additive manufacturing, the filler wire input is independent from the energy input and, for this reason, the parameters for material input and energy can only be varied independently to a limited extent. A proper height alignment of the wire tip with the electron heat source is necessary for the material transfer and a stable melt pool [10]. Since the melt pool and melt pool temperature can be controlled by process parameters [11–13], and weld beads are formed during solidification, knowledge about the correlation between process parameters and weld bead geometry is essential for process control and to understand the fundamentals of EBAM. Starting from weld bead geometry and knowledge of single bead profiles opens the capability for finding an optimum overlapping distance and is an important input for the subsequent welding sequence and to build up more complex geometries [14].

The mechanical properties during static and dynamic loading are influenced mainly by (1) the segregation of elements, (2) volume defects and (3) the microstructure [15,16]. During additive manufacturing, defects are mainly related to porosity affected by the process parameters and the type of adding material [17–19], cracks formation due to residual stresses generated during the process and delamination due to incomplete melting [18]. Insufficient bonding—e.g., lack of fusion as the major defect—has a negative effect on the static and dynamic mechanical properties of AM parts [20,21]. The variety of microstructure has a significant role in the mechanical properties of components made by titanium alloys [22,23]. In the case of additive manufacturing processing, the microstructural features (prior β grains, martensite, and α phase morphology, among others) depend on the experienced thermal cycle provoked by the type of process. Zhang et al. [24] summarized the progress in powder-based EBM in the field of α + β and β titanium alloys for solid and porous components, describing not only the microstructure and the mechanical properties, but also the unfavorable issues in the process and complex thermal conditions. The accumulation of layers and multiple thermal cycles in prevailing vacuum atmosphere can lead to heat accumulation and low cooling rates in EBM [24]. SLM produce mostly martensitic structure due to the cooling rate higher than 1000 K/min [19,22,25,26] reached in the layering. This microstructure can be decomposed in two phases ($\alpha + \beta$) by post heat treatment [19] or by thermal cycling during layer-by-layer building (i.e., EBM [24,27–29]), obtaining fine lamellar, colony, or Widmanstätten α structure and small amount of β phase [15]. For processes with high deposition rate and energy, the cooling rates are lower and it can form directly $\alpha + \beta$ in basket weave Widmanstätten or lamellae α morphology depending on the previous build part (e.g., [19,30–35]).

Recent literature [3,4] is limited for electron beam additive manufacturing, where the emphasis is mainly on material properties and microstructural details, and not on the process and parameter selection itself. The present study aims to establish suitable parameters for the EBAM process in $\alpha + \beta$ titanium alloy Ti-6Al-4V and determine the influence of these parameters on the dimensions of the single and multi-tracks welds for building walls with an optimal sequence of building. Furthermore, the microstructure and residual stresses produced during the manufacturing are analyzed, and the mechanical properties are estimated for the as-deposited and stress relief heat treated conditions.

2. Materials and Methods

2.1. Materials

Titanium alloy wire AWS A5.16 ER Ti5 (EN ISO 24034) was used with a commercially available diameter of 1.2 mm. The deposition was carried out with two different substrates: Commercial pure Titanium Grade 2 for the parameter studies and the single bead experiments; and Titanium Grade 5 (Ti-6Al-4V) for the subsequent multilayer-experiments, and mechanical as well as metallographic characterization.

The chemical compositions of the applied titanium alloy wire and the AM structures made out of it were determined as follows: the proportion of the elements aluminum (Al), vanadium (V), and iron (Fe) are determined by atomic absorption spectroscopy. The interstitial elements carbon (C) via solid-state infrared absorption detection method with LECO CS230 (LECO Corporation, St. Joseph, MI, USA); nitrogen (N) via thermal conductivity detection method with LECO ON 736 (LECO Corporation, St. Joseph, MI, USA); and oxygen (O) via non-dispersive infrared absorption detection method with LECO TCH600 (LECO Corporation, St. Joseph, MI, USA). Table 1 shows the chemical composition of the applied wire in this work.

Table 1. Measured chemical composition of applied titanium alloy wire AWS A5.16 ER Ti5 (EN ISO 24034); n.m.: not measured.

Material	Al (wt. %)	V (wt. %)	Fe (wt. %)	Ti (wt. %)	C (wt. %)	N (wt. %)	O (wt. %)
Solid Wire	6.36	3.48	0.11	bal.	0.018	<0.005	n.m.

For the titanium alloy wire, the proportion of all elements with the exception of vanadium is according to the specification. Only the proportion of vanadium is slightly below the permissible lower limit of the specification.

2.2. Experimental Setup

All welding experiments were performed using the electron beam welding machine pro-beam EBG 45-150 k14 (Probeam GmbH & Co. KGaA, Gilching, Germany). The pressure at the working chamber was below 5×10^{-3} mbar. Inside the chamber, with a nominal volume of about 1.4 m^3, a three-axis working table was used to handle the substrate plate. The welding filler wire was fed by pressure rollers and guided through a polytetrafluoroethylene (PTFE) hose and ends in an in three-axis movable water-cooled wire nozzle. The angle between the electron beam and the fed wire was 55°, and the distance of the nozzle to the substrate surface was 8 mm. The selected alignment was a trade-off between a proper wire guidance and a certain distance to avoid overheating of the nozzle during the process. Furthermore, a height distance of 1 mm between the wire tip and the substrate was set to ensure a continuous liquid metal bridge and thus a stable metal transfer during the process.

The main process relevant input and beam oscillation parameters are listed in Table 2. The acceleration voltage U_{acc} was kept constant and the power input was adjusted by varying the beam current I_{beam}. In order to adapt the power density, a customized beam figure consisting of several concentric circles with defined radii and elements was applied. A maximum outer diameter of 4 mm was chosen. In addition, the welding speed v_{weld} (9.0–11.0 mm/s) as well as the wire feed rate v_{wire} (2.7–3.9 m/min) were altered for the design of experiment (DOE) approach.

Table 2. Summary of input and beam oscillation (bop) parameters for welding and adjustment of wire feed unit.

Attribute	Abbreviation	Unit	Values
Acceleration voltage	U_{acc}	kV	90
Beam current	I_{beam}	mA	17.5–46.7
Welding speed	v_{weld}	mm/s	9.0–11.0
Wire feed rate	V_{wire}	m/min	2.7–3.9
Feed angle	α_{feed}	-	55
Focal point	f_p	-	Substrate Surface
Beam figure (bop)	-	-	Concentric Circles
Frequency (bop)	f	Hz	1000
Amplitude of deflection (bop)	x,y	mm	Ø 4

Two full factorial DOE's (Table 3) were carried out in order to evaluate the parameter influence on the weld bead width and height of the deposited single beads. The analysis of the dilution was done with a fractional factorial design in the moderate power range (Table 3). The obtained data was statistically evaluated, analyzed, and visualized by the means of the software MiniTab 19 (Version 19.1, Minitab, LLC, State College, PA, USA).

Table 3. Factor levels and combinations of full factorial design (FFD); moderate and higher values of deposition rates; and power input.

Factor		Unit	Heat Input			
			Moderate		High	
			Low	High	Low	High
Beam current	I_{beam}	mA	17.5	21.4	31.1	46.7
Welding speed	v_{weld}	mm/s	9.0	11.0	9.0	11.0
Wire feed rate	V_{wire}	m/min	2.7	3.3	3.3	3.9

2.3. Heat Treatment

The conducted heat treatment was selected according to literature. The post weld heat treatment (PWHT) has the purpose of soft-annealing and stress relieving. It was done in the preheated furnace at atmosphere for 2 h at 710 °C [15,36,37] and cooled in still air. The formation of alpha case during the heat treatment was irrelevant and any indication of oxidation at the surface was removed for subsequent mechanical characterization.

2.4. Metallographic Characterization

Macroscopic analysis was conducted by the means of a stereo microscopy (Zeiss Discovery V20, Zeiss, Oberkochen, Germany) to observe the strategy of building and defects related to the process (lack of fusion, pores, etc.). Microstructure and the effect of the process parameter on the dilution of single beads and overlapping distance for multi-track building was analyzed by a light optical microscope (Zeiss Axio Observer Inverted, Zeiss, Oberkochen, Germany). The geometry of the weld beads was measured by the related image processing program AxioVision 4.8.2 (Zeiss, Oberkochen, Germany).

Field emission scanning electron microscopy (SEM, TESCAN Mira 3, Tescan, Brno, Czech Republic) was used for characterization of the final microstructure after processing and heat treatment.

The examined embedded cross-sections were immersed in a Kroll's reagent etchant with a solution of 2 mL hydrofluoric acid (40%), 4 mL nitric acid (65%), and 94 mL distilled water. The etching duration was in the range of 3–40 s, observing a faster effect of the etchant on heat treated samples.

2.5. Mechanical Characterization

The mechanical properties of the AM structures were evaluated by tensile tests, Charpy V-notch impact tests, and hardness measurements at room temperature (20 °C).

The tensile testing specimens (DIN 50125-C, 6 mm diameter × 30 mm gauge length) were machined out of the block structures and are orientated along the welding direction. Therefore, the tensile properties were determined only in the longitudinal direction for the as-deposited and PWHT-condition due to the limited volume of processed material. The tensile tests were performed according to DIN EN ISO 6892-1 [38], with a constant testing speed of 1 mm/min on tensile testing device Zwick RMC 100 (ZwickRoell, Ulm, Germany).

Charpy V-notch impact tests were carried out according to DIN EN ISO 148-1:2017 [39]. The impact energy was measured for a longitudinal and transversal oriented samples in relation to the welding direction for the as-deposited and PWHT-condition (Figure 1a). The V-notch is located perpendicular to the sample orientation and the recorded absorbed energy represents the actual impact energy rectangular to the actual orientation. Therefore, the longitudinal orientated sample represents the properties across the welding direction, whereas the transverse sample represents the properties along the welding direction.

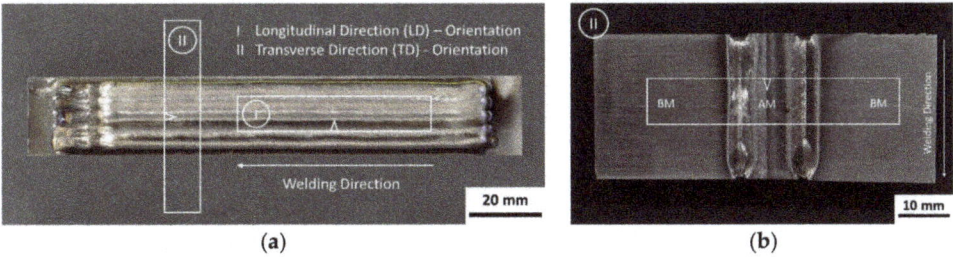

Figure 1. Charpy V-notch specimens: (**a**) orientation of the specimens in the AM blocks; (**b**) electron beam welded transverse Charpy V-notch specimen.

To determine the impact toughness along the welding direction according to the DIN EN ISO 148-1:2017 [39] standard and required dimensions, a novel approach was chosen. A trimmed transverse AM inlay was electron beam welded to parent material cantilevers (Figure 1b). The characteristic high energy density provided by EBW process results in a narrow weld zone. It might be assumed that the actual fracture area and the related recorded impact energy are not influenced by the previous EB welds.

Vickers hardness measurements according to DIN EN ISO 6507-1:2016 [40] were performed by using a load of 0.5 kgf (4.903 N) and a dwell time of 15 s (HV 0.5). The measurements were carried out on EMCO M1C hardness testing device (EMCO-Test, Kuchl, Austria). The hardness distribution of the AM cross-sections was obtained by hardness mapping visualized by the software Origin (Version 8.6, OriginLab Corporation, Northampton, MA, USA) and line scans for the as-deposited and PWHT conditions.

2.6. Residual Stress Measurements in Single Wall

The residual stress measurements in one single-track multilayer EBAM wall were carried out by energy dispersive X-ray diffraction (EDXRD) on the high-energy beam line ID15A at the European Synchrotron Research Facility (ESRF) in Grenoble, France. The characterization used a high-flux white beam with an energy range of 50–250 keV (wavelength range of 0.2480–0.0496 Å). Measurements were done in transmission mode with a slit size of 100×100 µm^2, giving a diamond shaped gauge volume with a length of approximately 2 mm. Diffraction spectra were collected by two energy-discriminating detectors placed at diffraction angles of $2\theta = 5°$ in the horizontal and vertical direction, allowing determination of strains in two directions simultaneously. Three parallel lines at the middle of the wall length were measured, spaced 2 mm apart to obtain a good reproducibility. A total of 10 points per line were acquired from the bottom (interface between the wall and the substrate) to the upper part with 1.6 mm space between the points. Each point was measured for 60 s, and during the measurement the sample was moved back and forth ±1 mm to increase statistics. Pawley refinement using GSAS (General Structure Analysis System) [41] was used to extract the lattice parameters (a and c) of the α phase in both directions. The low volume fraction and strong texture prevented the lattice parameter of the β phase to be determined reliably. The same AM build was measured before and after PWHT for 2 h at 710 °C.

3. Results

3.1. Single Layer Experiments

3.1.1. Dimension of Single Beads

The relation of the process parameters on the geometric evolution of the deposited weld beads by two full factorials DOEs is illustrated in Figure 2. The main effect plot visualizes the consequence of the predefined factors I_{beam}, v_{weld}, and v_{wire} on the target values: (1) weld bead width b and (2) weld bead height h. Considering the analyses' main effects on magnitude and slope, the single strength of the effects for the selected input factors of the moderate and the high input DOE show a good agreement with the literature [9]. The steeper the slope of the line, the greater is the impact of the factor on the geometry of the single beads. For both type of input (factorials), the increment of the I_{beam} increases the width of the bead with a slight change in the height, while the increase of the v_{weld} produced lower and narrower weld beads due to reduced energy input and material input per unit length. The wire feed rate v_{wire} has a strong influence on the weld bead height during processing at moderate input. The bead height increases by increasing the v_{wire} with a negligible change in the width (nearly horizontal line in Figure 2a). For high input, there are not significant changes of the bead geometry by varying the v_{wire}.

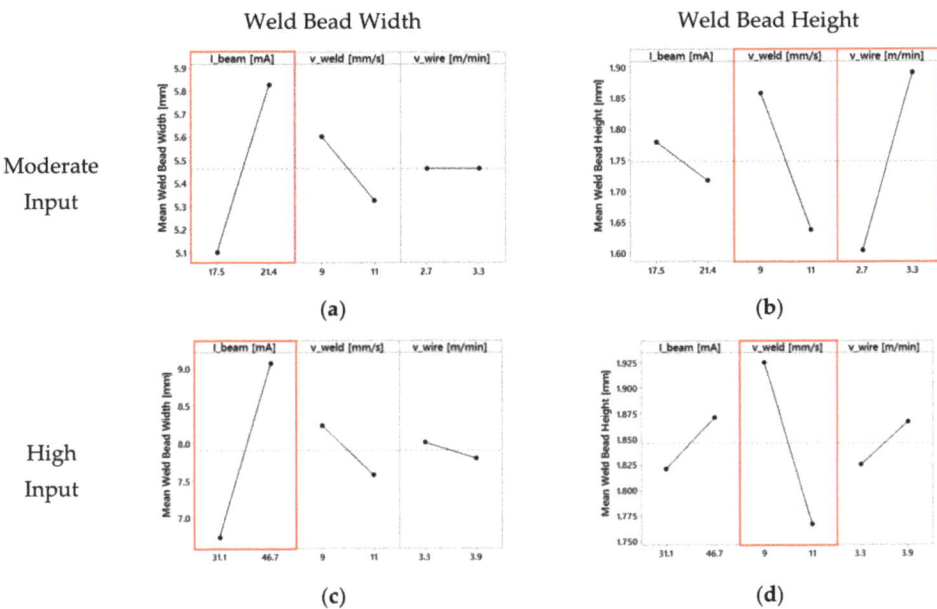

Moderate Input

High Input

Figure 2. DOE analysis: Influence of process parameters on the weld bead geometry at (**a**,**b**) moderate and (**c**,**d**) high input, and related significant process parameters are framed red.

3.1.2. Dilution of Single Bead

The weld bead dilution D is defined by the ratio between the molten base material (BM) and the whole weld bead cross section [42]. A minimum dilution leads to produce an efficient and fast build-up, and it would be reached by low energy per unit length and high deposition rate. In this work, the parameters analyzed by DOE for moderate heat input (Figure 2a,b; Table 3) are used for the dilution estimation. The analysis of the dilution was done with a fractional factorial design, and Table 4 summarizes the parameter used for the single bead. The ratio of wire feed rate to welding speed represents the material input per length (v_{wire}/v_{weld}) and amount of weld bead reinforcement. Figure 3 shows the cross section and dilution obtained for each parameter.

Table 4. Database and settings to Figure 3.

Figure 3	Beam Current (mA)	Welding Speed (mm/s)	Wire Feed Rate (m/min)	Material Input per Length (-)	Dilution (%)
(a)	17.5	11	2.7	4.1	45
(b)	21.4	11	3.3	5.0	45
(c)	21.4	9	2.7	5.0	52
(d)	17.5	9	3.3	6.1	28

Figure 3a,b shows the same dilution of 45% for different applied process parameters. Therefore, different process parameter configurations may be possible to achieve the same dilution, which should be minimized.

From Figure 3b to Figure 3c, not only the welding speed is reduced from 11 to 9 mm/s, but also the wire feed rate is reduced from 3.3 to 2.7 m/min. It results in a constant ratio of wire feed rate to welding speed, which corresponds to a constant material input per length value of 5.0. Since the welding speed (energy input per length) has a weak influence on the weld bead width (Figure 2a,c), more base material is fused by decreasing the welding speed from 11 to 9 mm/s (Figure 3b,c). By increasing the area of fused base material and keeping the material input per length constant, the dilution consequently increases from 45% to 52%.

A reduction of the beam current from 21.4 mA (Figure 3c) to 17.5 mA (Figure 3d) results in less area of the molten base material. By simultaneously increasing the wire feed rate, hence material input per length, the dilution can be optimized to a minimum value. Observations show a minimum dilution of 28%, and this particular configuration of the parameters is used for subsequent experiments.

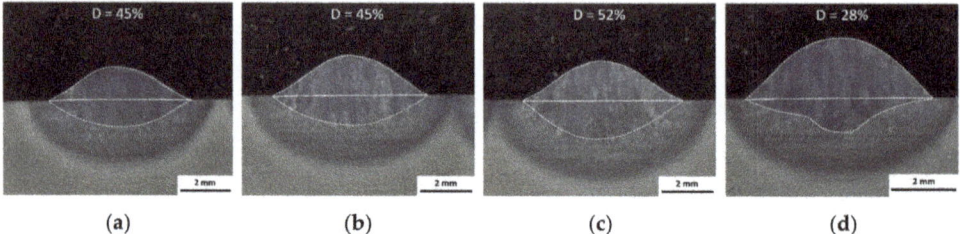

Figure 3. Dilution and geometric evolution of single beads deposited with different parameter configurations given in Table 4. Dilution: (**a**,**b**) 45%, (**c**) 52% and (**d**) 28%.

3.1.3. Overlapping Distance for Multi Track

The production of structures by EBAM requires the optimization of the overlap of the beads for a uniform and flat surface for successive layers during the build-up process. The overlapping distance given by the axial offset (d) is linked to the width (W) of the single bead, as shown in Figure 4a. Figure 4b–f show for different ratios (d/W) the form and symmetry of the deposited overlapped beads. If the overlapping distance is too small (e.g., ratio 0.55) an asymmetric layer is observed. The increment of the distance helps for the symmetry in the overlapping, with even surface up to ratio of 0.75, where wavelike surface with valleys in between individual beads is observed (Figure 4f). A compromise overlapping distance was identified in the range of 70–75% of the bead-width, and is in good agreement with Suryakumar et al. (66.6%) [43] or Ding et al. (73.8%) [44] reported for gas metal arc welding (GMAW) process. The manufacturing of the AM structures in the following sections is done with an axial offset of 4 mm between the deposited tracks, which means an overlapping distance of about 72%.

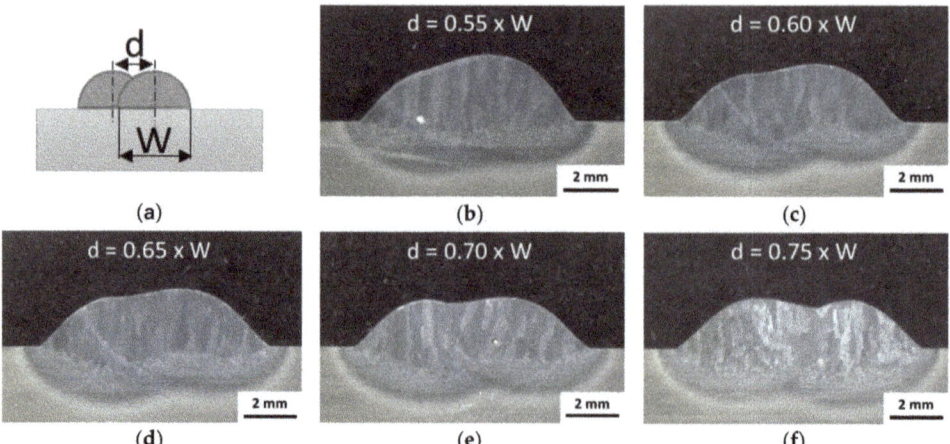

Figure 4. Multi-track overlapping distances (ratio d/W): (**a**) scheme of the overlapping with the main parameters d and W, (**b**) 0.55, (**c**) 0.60, (**d**) 0.65, (**e**) 0.70, and (**f**) 0.75.

3.2. Building AM Block

AM blocks were built by multi-track and multi-layer production, using optimized parameters in Section 3.1. The power input was steadily decreased (approx. 0.5 mA after every second layer) by increasing the number of layers to compensate the preheating effect of prior layers and the consequently reduced heat flux with increasing height. Figure 5a shows the transverse cross-section of the built AM block after seven layers with a linear growth rate of the height by adding continuously layers (Figure 5b). The derived linear fit indicates an average growth rate of about 1.64 mm for each layer that helps to set up an automated process without needs of venting the chamber.

According to literature [14,32,45–51], different welding sequences are reported and well established for AM processes. An alternating symmetrical welding sequence permits a robust process design for rectangular structures with parallel walls (Figures 5a and 6a). For that, the first bead must be welded in the center of the previous layer and the subsequent beads follow from inside out, alternating on both sides (see marks in Figure 5a).

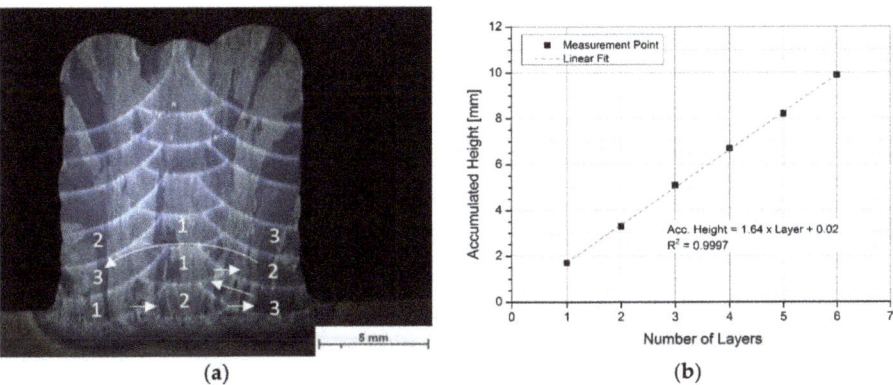

Figure 5. (a) Alternating symmetric welding sequence to build-up rectangular AM structures and (b) measured accumulated height over number of layers and ascertain growth rate by the means of linear fit.

3.3. Chemical Composition AM Block

The chemical compositions of the studied AM blocks are given in Table 5. The chemical composition of the AM material shows only evaporation losses of aluminium in the range of approx. 1 wt % (approx. 14%). The measured oxygen concentration of 0.11 wt % is within the specification and indicates no further oxygen pick-up by the ambient vacuum atmosphere or contamination.

Table 5. Chemical composition of AM bulk material.

Material	Al (wt. %)	V (wt. %)	Fe (wt. %)	Ti (wt. %)	C (wt. %)	N (wt. %)	O (wt. %)
AM block no. 1	5.47	3.39	0.11	bal.	0.017	<0.005	0.11
AM block no. 2	5.49	3.46	0.10	bal.	0.019	<0.005	0.11

Since aluminum is the element in Ti-6Al-4V with the highest saturated vapor pressure, it can be expected that it has a significant tendency to evaporation and vaporization loss during processing. The evaporation of aluminum by processing Ti-6Al-4V with different AM processes has already been reported in the literature (i.e., EBF^3 [4,52], E-PBF [53,54], or L-PBF [55]). Unlike L-PBF, the excessive aluminum loss was reported for vacuum based processes E-PBF (up to 30%) and EBF^3 (up to 39%) by Juechter et al. [54] and Xu et al., respectively [4]. By minimizing the dilution, the excessive re-melting of previous layers and additional evaporation of aluminum can be avoided. The evaporation loss of aluminum in the present study is significantly lower compared to the investigations of Xu et al. [4].

3.4. Metallography

Metallographic investigations were performed on cross-sections of AM blocks built by 10 layers × 5 beads, with an alternating symmetric welding sequence (Figure 6a).

The AM cross-section shows epitaxial grown columnar prior β-grains, reaching over several layers and parallel to the build direction. The form and size of the columnar prior β-grains is provoked by the large temperature gradient during the solidification in direction of the heat flow [56,57]. A very low amount of macro and micro pores with a random distribution is observed along the section. Layer bands (Figure 6a,c) represent a minimal change of the microstructure (HAZ) due to the thermal cycles and intrinsic heat treatment of the neighbored weld bead deposition during the building process, as observed as a dark zone by etching effect in wire based investigations [33,56,58,59]. Recent work has demonstrated a minimal segregation near to these layers, mainly related to β-stabilizer elements [59]. The transition between the AM-Structure and the substrate consists in coarser equiaxed prior β-grains (Figure 6d).

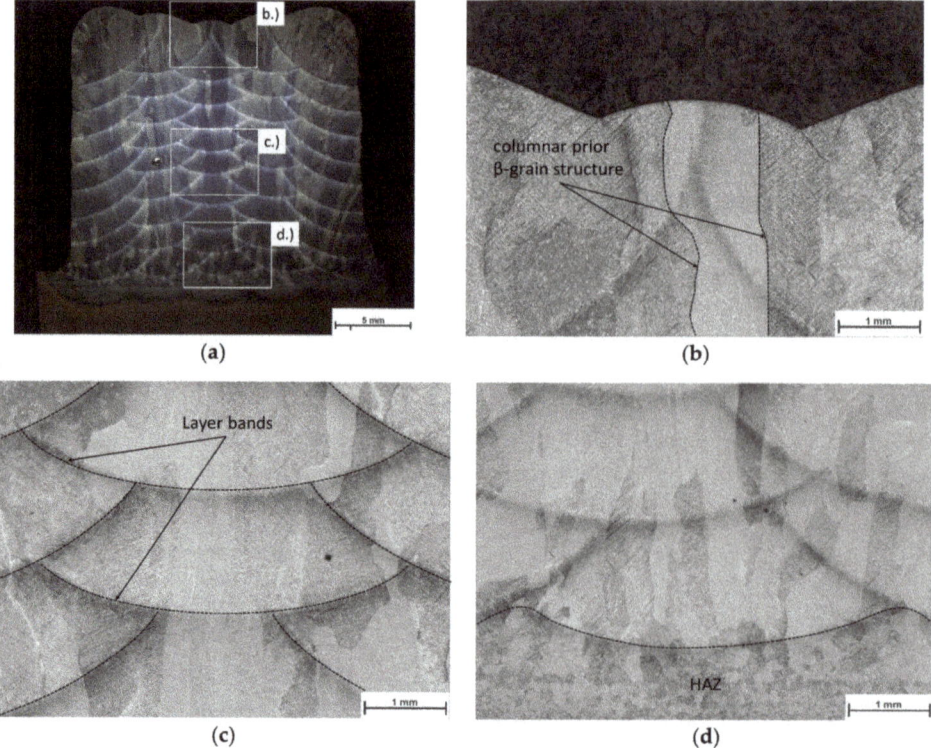

Figure 6. (a) Macrostructure with layer bands and columnar structure, (b) columnar prior β-grain structure, (c) presence of several layer bands, and (d) transition AM bulk material to heat-affected zone (HAZ).

The fast cooling rates reached during the process (see Supplementary Data, Figures S1 and S2) promote the formation of a mixture microstructure of finer α and martensite (α') along the whole AM structure, as shown in Figure 7a,c. The morphology of the microstructure is almost homogenous in the whole building block. After the PWHT, the microstructure decomposes into in α + β structure, with a fine α-lamellar structure (Figure 7b). Precipitation of fine β within the fine α lamella detected in the Figure 7d can be probably due to an heterogeneous enrichment of β-stabilizer elements [60].

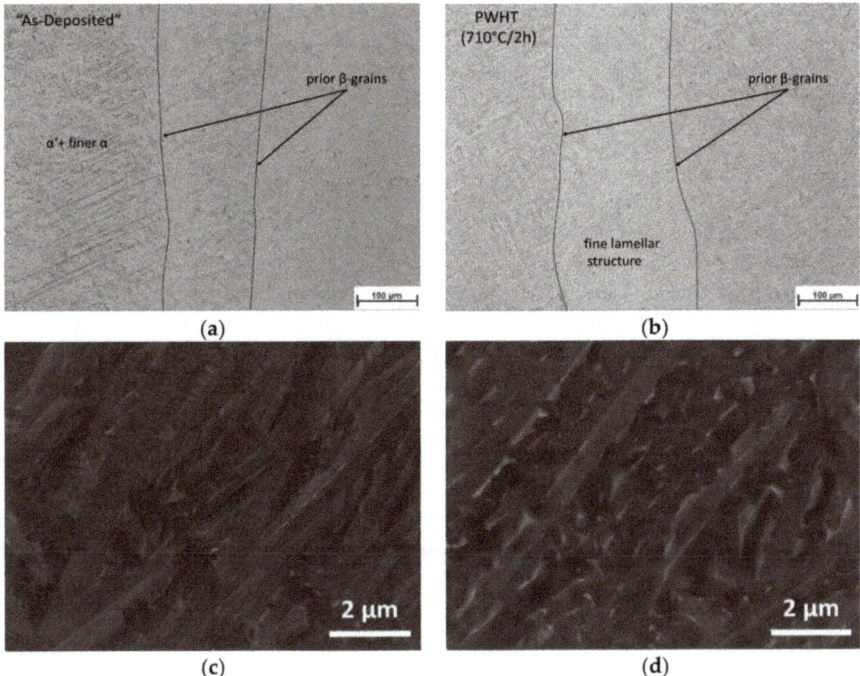

Figure 7. Microstructure of the AM block center: (**a**) as-deposited and (**b**) after PWHT condition; Detail of the microstructure by SEM investigations in (**c**) as-deposited and (**d**) after PWHT conditions.

3.5. Residual Stresses by High Energy Dispersive X-ray Diffraction

Diffraction spectra in both directions for the three different linescans is illustrated in Figure 8a as 2D-plot (position vs. d-spacing) for as-deposited condition. The combination of large grains, strong texture and parallel X-ray radiation results on variations across the height, where certain peaks appear and disappear in in the spectra. This resulted in d-spacing errors in the range of $(50–100) \times 10^{-6}$ and $(100–150) \times 10^{-6}$ for a and c lattice parameters, respectively, from refinements. In the following, results from all three linescans were averaged.

The information obtained by both detectors helps to determine the spatially resolved residual strains and stresses. Strains are calculated as function of the lattice parameter a and c [61]

$$\varepsilon = (2 \times \varepsilon_a + \varepsilon_c)/3, \tag{1}$$

where,

$$\varepsilon_a = a/a_0 - 1, \tag{2}$$

$$\varepsilon_c = c/c_0 - 1, \tag{3}$$

Reference lattice parameters are calculated from the average of the values measured in the vertical direction in the PWHT condition due to the constant value observed in all the positions. a_0 is 2.92612 Å and c_0 is 4.67067.

Figure 8b shows the calculated residual strains in the x- and y-direction for the as-deposited sample. Generally, the all three principal strain need to be considered in order to calculate the corresponding stresses, which requires measurements of strains also in the z-direction. However, as seen in Figure 8b there is an excellent agreement between the measured ε_y and the values calculated as—$\nu \cdot \varepsilon_x$ based on the assumption of a uniaxial stress in the x-direction (where $\nu = 0.317$ is Poisson's ratio [21]). Consequently, only strain and stresses in the x-direction are considered hereafter. Also included in Figure 8b is ε_x for the PWHT condition, which shows significantly slower strain levels.

Under the assumption of a uniaxial stress state, the stress–strain relationships simplify to

$$\sigma_x = E \times \varepsilon_x, \qquad (4)$$

and $\sigma_y = \sigma_z = 0$. The Young's modulus, E, for this alloy is 120.4 GPa [21]. Figure 8c shows the resulting residual stresses as a function of height in the wall. The error bars represent the scatter obtained from averaging the results from the individual linescans at each position. Residual stresses lie in the range of 200–450 MPa in tension for the x-direction, which is a result of the thermal contraction of the deposited material. The trends agree well with previous measurements and process simulations of AM Ti-6Al-4V walls [62]. After annealing, the stresses are lower than 100 MPa and closer to zero in the upper part of the single wall.

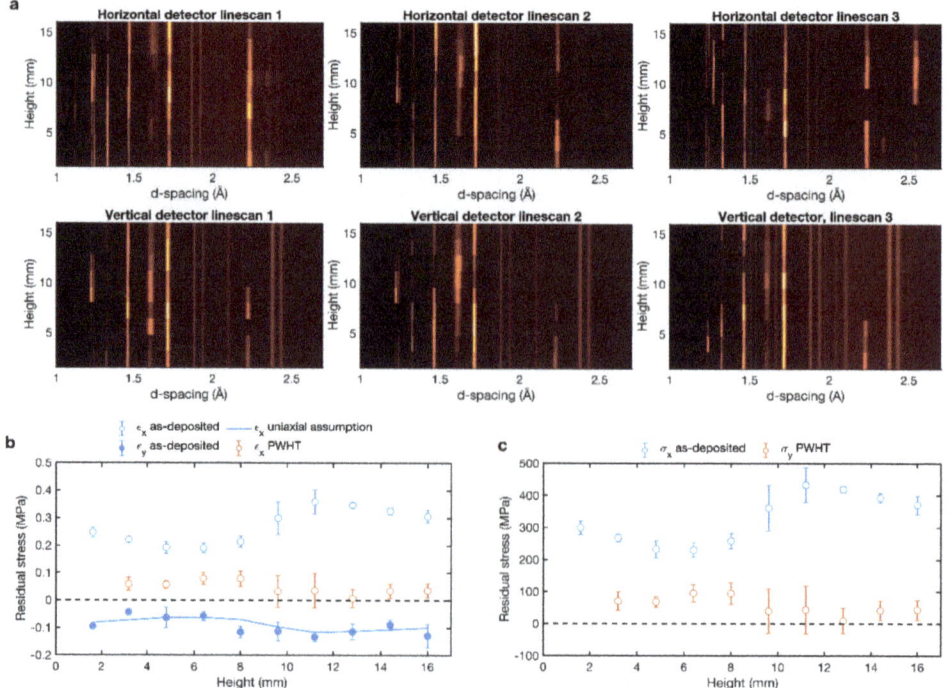

Figure 8. (a) 2D plot diffraction spectra of the as-deposited condition, for the horizontal and vertical detectors as a function of height in the wall. (b) Residual strains and (c) residual stresses for as-deposited and PWHT conditions as a function of height in the single wall.

3.6. Mechanical Properties

3.6.1. Tensile Test

The tensile stress over strain curves for the longitudinal orientated specimens in the as-deposited and PWHT condition are presented in Figure 9a. The as-deposited condition has a yield tensile strength (YS) of 846 MPa, ultimate tensile strength (UTS) of 953 MPa, and a fracture elongation (El) of 4.5%. When the heat treatment (PWHT) is carried out, a slight reduction of the YS and UTS values is observed but the El increases by more than 100% up to 9.5%.

According to the specifications for wrought material Ti-6Al-4V (YS > 795 MPa, UTS > 860 MPa, and El > 10%) [22,63], the tensile strength of the AM block structure in as-deposited condition is on the lower limit of the specification limit and the low fracture elongation might be related to the type of cast material. The fractured surface (Figure 9b) shows a ductile morphology, denoted by transgranular dimples and microcracks formed by coalescence of voids. This observation is similar to materials made by WAAM process [34]. The nucleation of voids is observed at the α/α' and β phases interface, and they start to grow and coalesce along with the interface. Example of voids and microcracks are illustrated in Figure 9c and d for as deposited and PWHT conditions, respectively.

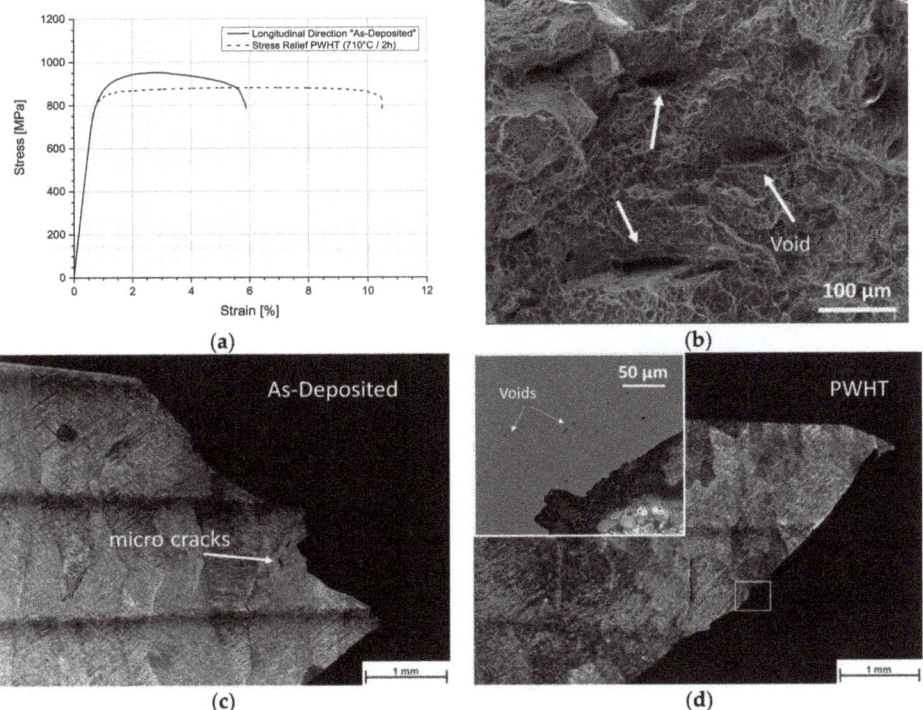

Figure 9. (a) Engineering stress–strain curves for longitudinal orientated tensile specimens. (b) Detail of the fracture surface for the PWHT condition. (c,d) Macrograph of the cross section of the fracture area of as-deposited and PWHT conditions, respectively.

3.6.2. Charpy V-Notch Impact Tests

The absorbed energy and measured lateral expansion of the tested specimens are listed in Table 6. The results show generally a high impact toughness in the as-deposited condition in comparison to experiments carried out by SLM or EBM [64–66] and a slight reduction of the absorbed energy after PWHT. The orientation of the sample affected the impact energy. Specimens with longitudinal (LD) orientation shows higher value than the transverse oriented (TD) samples. Furthermore, a non-isotropic behavior is observed and it might be related to the imperfections produced during the process, as exemplified in Figures 10 and 11.

Table 6. Results of Charpy V-notch impact testing in longitudinal (LD) and transverse direction (TD) for as-deposited and PWHT condition.

Sample Orientation	As-Deposited		PWHT	
	Impact Energy (J)	Lateral Expansion (mm)	Impact Energy (J)	Lateral Expansion (mm)
LD	54	0.6	45	0.4
TD	40	0.3	34	0.3

Stereo micrographs in Figure 10 show irregularities in the fractured surfaces. The fractured surfaces of the LD samples are characterized by randomly distributed and scattered imperfections, mainly related to macro (about 200 µm size) and micro (approx. 5 µm) porous and sharp-edged cavities (Figure 11a). In the case of TD samples, uniform and recurring imperfections in the direction of welding for the entire weld length might be related to a lack of fusion during the process, as observed in Figure 11b.

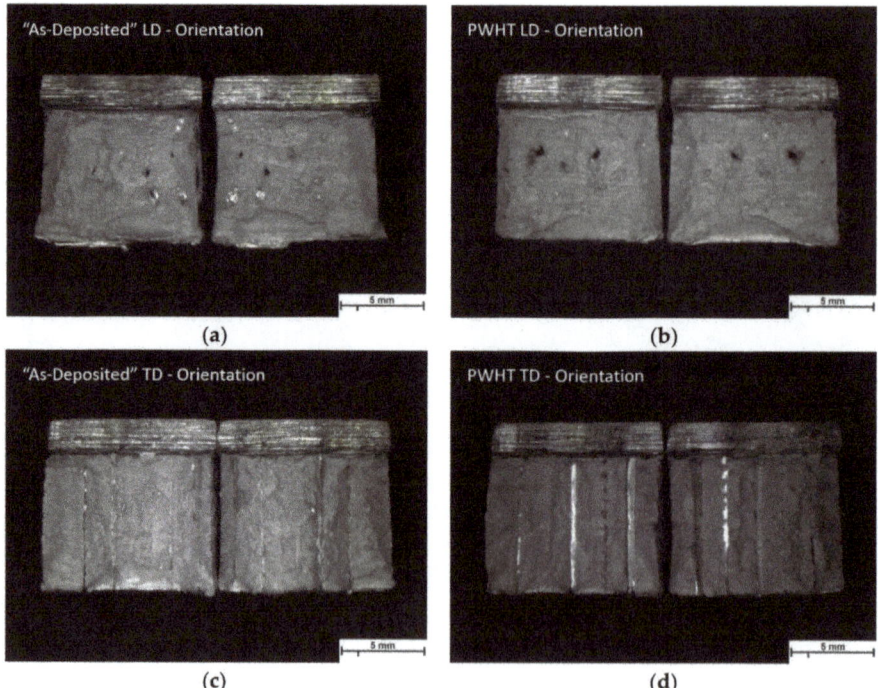

Figure 10. Stereo macrographs of the fractured surfaces in the (a,c) as-deposited and (b,d) PWHT condition of the tested longitudinal and transverse orientated Charpy V-notch specimens.

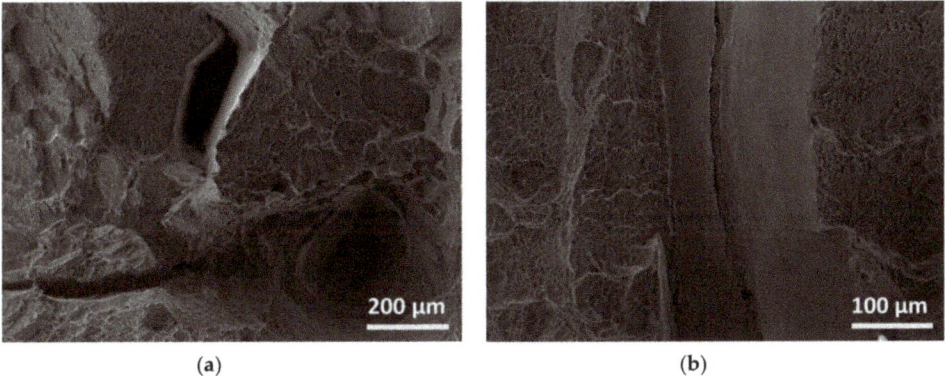

Figure 11. SEM micrographs of the fractured surfaces of Charpy V-notch specimens: (**a**) as-deposited in LD-orientation and (**b**) PWHT in TD-orientation specimens.

3.6.3. Hardness Testing

The hardness mapping of the as-deposited condition (Figure 12a) shows an almost uniform distribution of the hardness in the whole transverse cross-section, with an average value of 334 ± 9 HV. The hardness decreases slightly at the fusion lines and the heat-affected zone (HAZ) between the AM block and substrate shows the highest values of hardness (~400 HV). Similar distribution of hardness is observed in PWHT condition, represented in Figure 12b by a line scan from top of the building block to the substrate. The average value for the heat-treated condition is 329 ± 9 HV.

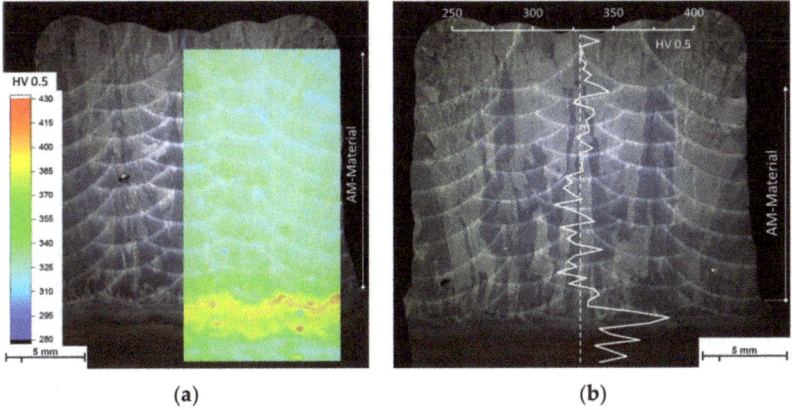

Figure 12. (**a**) Hardness-mapping of transverse cross-section in as-deposited condition and (**b**) hardness line scan in PWHT condition (standard deviation = ±9 HV).

4. Discussion

4.1. Electron Beam Processing

4.1.1. Dimension of Single Beads

The single track experiments showed that the width and height of the welding beads can be adjusted by the machine parameters. The evaluated parameters for the low input DOE have nearly the same tendencies as Wallace et al. described [9]. Wide beads can be achieved especially at high current, while high beads can be produced by a big amount of wire per unit length.

Sliva et al. [67], Gudenko et al. [68], and Dragunov et al. [69] pointed out the importance of beam oscillation parameters on the weld bead formation. By using concentric beam oscillations, the change of oscillation parameters (e.g., amplitude) together with beam current, the electron beam energy distribution and thus the distribution of energy input in the active zone can be adjusted. In the present study, for all experiments, the beam oscillation parameters (Table 2)—i.e., electron beam distribution—were kept constant on a circular area of a diameter of 4 mm and only the increase of the beam current results in an increase of the energy input by a similar beam distribution. Partial melting of the substrate by the electron beam is required to form the desirable liquid metal bridge material transfer mode between substrate and wire [70,71]. Tang et al. [13] investigated the heat transfer in EBF^3 and simulated fluid flow of the weld pool by a 3D transient model. In EBF^3 the high temperature region is restricted to the area of a direct electron beam exposure, showing a steep temperature gradient to its surrounding. An enlargement of the fusion width predicted by simulation of Tang et al. [13] shows a directly proportional relationship to the increase of the beam current, in good agreement with the present study. Taking literature and present study into account, it can be concluded that the melt pool width and therefore maximum weld bead width is predominantly described by beam oscillation parameters (e.g., area of beam exposure i.e., amplitude of beam deflection) as well as the selected beam current. In contrast to the width, the weld bead height is mainly determined by the wire feed rate and the welding speed. For gas metal arc welding (GMAW) process, Xiong et al. [72] observed that the ratio of wire feed rate to welding speed mainly influences the shape of the weld bead profile. Independent of the applied wire-based AM process, the ratio describes the material input per unit length. In EBF^3 no distinct evaporation of titanium (predominantly evaporation of aluminum [4,52]) and no larger material loss is expected, therefore the composition of the melt pool consists out of the molten substrate/previous layer and fed filler wire. The volume of weld bead reinforcement is therefore determined by the amount of material input per unit length and the ratio of wire feed rate and welding speed. As the weld bead width is not significantly influenced by the velocities, any change of wire feed rate or welding speed (i.e., material input per length) results in a direct change of the weld bead height.

4.1.2. Dilution of Single Bead

In AM, it is desirable to reduce the dilution to improve the efficiency of the stacking process and to avoid a large number of layers and extensive heat input. Within the examined process limits (Table 4), in principle, the lowest dilutions were realized by minimum beam current. A reduction of the beam current not only narrows directly proportional the fusion width, but also the fusion depth and thus the volume of the molten substrate/previous layer [13]. By minimizing the beam current to a threshold value and simultaneously increasing the material input per unit length—i.e., increasing the ratio of wire feed speed to welding speed—a maximum weld bead reinforcement and minimum dilution can be achieved. Though, a minimum melting of the substrate/previous layer is still mandatory for desirable material transfer mode [70,71]. The dilution of 28% in this work was comparatively low since the aim was to achieve a fast build-up process with minimum energy input. However, a minimal dilution and a certain amount of fused metal are required for integrity and for a compensation of wavelike surfaces with valleys between the individual beads of the previous layers. The selected process parameters and overlapping distances enabled a rapid stacking process at a low dilution, whereby bonding defects could not be completely avoided (e.g., Figure 11). It can be assumed that an adjustment of the parameters and avoidance of lack of fusion can further improve mechanical properties.

4.1.3. Overlapping Distance for Multi Track

Since more volumetric AM structures require several tracks per layer and independent of the feed stock, the overlap distance between the tracks is of particular interest for PBF (e.g., [73,74]), but also in wire-based processes (e.g., [10,14]). For arc welding processes there are several models in the literature for describing an optimized overlap in general [43,72,75–79]. The models are based on different theoretical assumptions, which mainly relate to the geometry of the weld bead and related cross-sectional profile. Even when recent models also take the spreading of the weld bead into previously deposited bead and thus changing the geometry, the transferability on weld deposits produced by EBF^3 is questionable. In EBF^3, unlike GMAW processes, the material- and energy input is decoupled and material transfer modes, but also heat source models, differ.

To the best of the authors' knowledge, no general models or recommendations for the overlap distance are proposed in the literature for wire-based additive manufacturing using electron beam so far. For this reason, the present study opted for an experimental approach, where the axial offset of the following track from the adjacent track was varied between 0.55 and 0.75 of the single track width. It produces a surface flatness in conjunction with an acceptable valley formation.

4.1.4. Building AM Block

Since the process takes place in a vacuum atmosphere, the heat is slower dissipated by heat conduction and thermal radiation, without the contribution of convection like in WAAM. The transient heat flux is distorted by the increasing number of layers, and the ratio of heat dissipated by thermal radiation to heat conduction increases. Since the temperature increases, weld bead geometry changes and tends to widen and flatten [80]. The changing weld bead height and deviation from the nominal values accumulate layer by layer. In wire-based AM processes (e.g., EBAM) where energy and material input are separated, the exact alignment of the filler material to the previous layer/substrate determines the material transfer mode as predominant [70,71]. If the relative position of the wire tip to the previous layer/substrate increases when increasing the number of layers, the material transfer mode changes from a preferable liquid metal bridge transfer (Figure 13b) to a droplet transfer (Figure 13a), promoting spatter formation and irregular metal deposition [10].

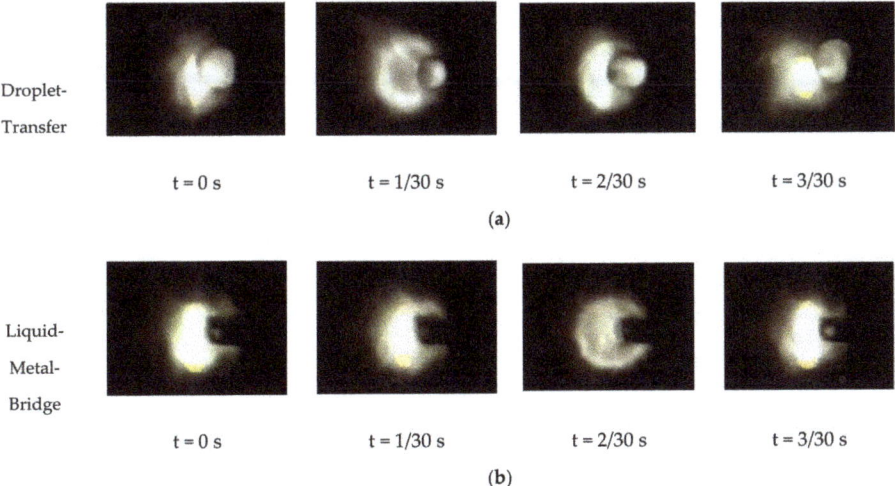

Figure 13. Material transfer modes in electron beam additive manufacturing recorded via CCT camera: (a) droplet transfer and (b) liquid metal bridge transfer.

By continuously reducing the beam current (i.e., power input) every second layer, the increasing preheating effect can be overcome, and a constant growth rate of 1.64 mm per new layer can be ensured. A uniform growth rate per layer is decisive for setting up an automated setup and guarantee stable material transfer.

The energy input and welding sequence per layer for Ti-6Al-4V affect the microstructure and related properties [81] due to the process' cooling rates. For gas metal arc welding, Stockinger et al. [50] and Plangger et al. [51] showed that an alternating welding sequence results in a uniform AM structure with homogenous properties for martensitic steels in transverse cross-sections. A similar approach was chosen and by alternating weld bead sequence, the energy input and heat can be distributed more uniformly. The morphology of the microstructure was almost homogenous in the whole transverse cross-section and also no changes in hardness for the AM material could be observed (Figure 12).

4.2. Effect of the Microstructure on the Mechanical Properties

The tensile testing showed that the achieved results are in the low range of typical values for this alloy. A comparison to additive manufactured parts with other processes from literature is shown in Figure 14. This difference in the values is related mainly to the microstructure built during the process. In general, as deposited conditions of SLM components have shown the highest values of tensile strength due to the fully retained martensite [20,81]. Buildings by LENS with different energies show mainly α′ martensite structure mixed with some acicular α phase, while EBM powder-based process retained very fine α laths due to longer time exposure at 650–750 °C during the manufacturing [29]. The lower values observed by Edwards et al. [82] are related to a slow cooling during the process and partial stabilization of the α phase. The apparent martensitic form observed in this work (Figure 7) has been decomposed into α + β during the deposition. The building sequence induces by the heat input a short time annealing effect on the microstructure, reducing the hardness to values closer to the PWHT (Figure 14). Although moderate/high cooling rate is expected in this process (~150–400 °C/s, as reported in the Figure S2), the high energy input of this process and the bad thermal conductivity of Ti-6Al-4V leads to keep warm temperatures in the last deposited layer [30]. On the other hand, the presence of residual stresses in the as deposited condition, might be related not only to the volume distortion provoked during the cooling but also to the partial transformation of the apparent martensitic structure, in agreement with observations in laser metal deposited process [62]. The deposited material shows lower strength values than typical martensitic structure in wrought material, but with the same elongation [23,83].

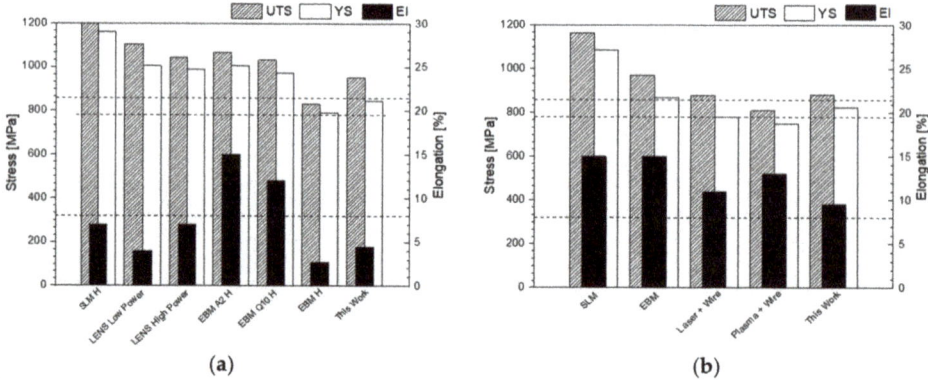

Figure 14. Comparison of the results obtained in the present work with other process. (a) As deposited condition is compared with values from LENS, ABM A2 H, and EBM Q 10 processes referred by Zhai et al. [29], SLM reviewed by Liu et al. [81] and EBM H by Edwards et al. [82]. (b) after PWHT is compared with a summary of processes given by Greitmeier [15] and Liu et al. [81].

Although there is no significant changes in the tensile strength in comparison with wire-feed processes [15] after stress relief treatment (PWHT), the values of strength are substantial lower than the observed in SLM process [81]. This difference might be related to the α lath size reached during the annealing treatment [81]. Therefore, further investigation to optimize the microstructure by thermal treatments is necessary to obtain the desirable mechanical properties.

5. Conclusions

In this work, the use of wire-based additive by electron beam process for Ti-6Al-4V was investigated. Suitable parameters for building single and robust walls were found and the microstructure in as deposited and after heat treatment was correlated with the mechanical properties. The findings can be summarized as follows:

- During the building of the wall, the bead shape is mainly affected by the beam current, the weld-velocity, and the feeding rate of the wire.
- A fast build-up process with minimum energy input guaranteed a dilution of 28%, which facilitate an optimal overlapping distance to reduce the wave-like surface.
- The use of a symmetric welding sequence with an overlapping distance of 70–75% of the bead-width permits a flat surface and a linear growth rate of the block.
- The chemical composition of the AM material shows only evaporation losses of aluminum in the range of approx. 1 wt %.
- As-deposited condition showed a mixture of finer α and martensitic structure within the coarser columnar prior β grains, providing a low tensile strength compared to similar additive manufacturing processes.

The lack of fusion observed during impact tests must be improved in further investigations for a precise relationship of microstructure and mechanical properties.

Supplementary Materials: The following is available online at http://www.mdpi.com/1996-1944/13/15/3310/s1, Figure S1: (a) temperature distribution for a single track deposition in electron beam additive manufacturing and (b) derived cooling rates, Figure S2: (a) exponential fit of the measured values and (b) derived cooling rates for a single track deposition via electron beam additive manufacturing.

Author Contributions: Conceptualization, F.P. and N.E.; Methodology, P.P., F.P., F.W., N.E., A.S., M.H.C., and R.P.; Investigation, P.P., F.P., F.W., N.E., A.S., M.H.C., and R.P.; Resources, N.E.; Writing—Original draft preparation, P.P., F.P., and F.W.; Writing—Review and editing, F.P., F.W., N.E., A.S., M.H.C., and R.P.; Supervision, N.E.; Project administration, F.P. and N.E.; Funding acquisition, N.E. All authors have read and agreed to the published version of the manuscript.

Funding: This research was funded by "Dobeneck-Technologie-Stiftung" and the COMET program within the K2 Center "Integrated Computational Material, Process and Product Engineering (IC-MPPE)" (Project No 859480). This program is supported by the Austrian Federal Ministries for Transport, Innovation and Technology (BMVIT) and for Digital and Economic Affairs (BMDW), represented by the Austrian research funding association (FFG), and the federal states of Styria, Upper Austria, and Tyrol. Open Access Funding by the Graz University of Technology.

Acknowledgments: The authors would like to acknowledge ESRF, experiment no. MA/3395, for granting the beam time in beamline ID15A for conducting the residual stress measurements presented in this work. The experimental support by the beamline responsible scientist Thomas Buslaps is also highly appreciated. Authors would thank to Hermith GmbH for the provision of substrate material for the experiments and Leander Herbitschek for the technical support during the process of AM parts. P.P. thanks to Wirtschaftskammer Steiermark (WKO) for the scholarship grant during his research study.

Conflicts of Interest: The authors declare no conflict of interest.

References

1. ISO 17296-2:2015. *Additive Manufacturing—General Principles—Part 2: Overview of Process Categories and Feedstock*; ISO: Geneva, Switzerland, 2015.
2. Taminger, K.; Hafley, R. Electron beam freeform fabrication: A rapid metal deposition process. In Proceedings of the 3rd Annual Automotive Composites Conference, Troy, MI, USA, 9–10 September 2003.
3. Wanjara, P.; Watanabe, K.; De Formanoir, C.; Yang, Q.; Bescond, C.; Godet, S.; Brochu, M.; Nezaki, K.; Gholipour, J.; Patnaik, P. Titanium alloy repair with wire-feed electron beam additive manufacturing technology. *Adv. Mater. Sci. Eng.* **2019**, *2019*. [CrossRef]
4. Xu, J.; Zhu, J.; Fan, J.; Zhou, Q.; Peng, Y.; Guo, S. Microstructure and mechanical properties of Ti-6Al-4V alloy fabricated using electron beam freeform fabrication. *Vacuum* **2019**, *167*, 364–373. [CrossRef]

5. Savchenko, N.L. Structure and phase composition of Ti-6Al-4V alloy made by additive manufacturing technology. *AIP Conf. Proc.* **2019**, *2167*, 020310. [CrossRef]
6. Taminger, K.; Hafley, R. Electron beam freeform fabrication for cost effective near-net shape manufacturing. In Proceedings of the NATO AVT-139 Specialists' Meeting—Cost Effective Manufacture via Net Shape Processing, Amsterdam, The Netherlands, 15 May 2006.
7. Schultz, H. *Electron Beam Welding*, 1st ed.; Elsevier Science & Technology: Cambridge, UK, 1994.
8. Sciaky Inc. Electron Beam Additive Manufacturing. Available online: http://www.sciaky.com/additive-manufacturing/electron-beam-additive-manufacturing-technology (accessed on 12 January 2020).
9. Wallace, T.; Bey, K.; Taminger, K.; Hafley, R. A design of experiments approach defining the relationships between processing and microstructure for Ti-6Al-4V. In Proceedings of the 15th Solid Freeform Fabrication Symposium, Austin, TX, USA, 2–4 August 2004.
10. Fuchs, J.; Schneider, C.; Enzinger, N. Wire-based additive manufacturing using an electron beam as heat source. *Weld. World* **2018**, *62*, 267–275. [CrossRef]
11. Gockel, J.; Beuth, J.; Taminger, K. Integrated control of solidification microstructure and melt pool dimensions in electron beam wire feed additive manufacturing of Ti-6Al-4V. *Addit. Manuf.* **2014**, *1*, 119–126. [CrossRef]
12. Gockel, J.; Fox, J.; Beuth, J.; Hafley, R. Integrated melt pool and microstructure control for Ti-6Al-4V thin wall additive manufacturing. *Mater. Sci. Technol.* **2015**, *31*, 912–916. [CrossRef]
13. Tang, Q.; Pang, S.; Chen, B.; Suo, H.; Zhou, J. A three dimensional transient model for heat transfer and fluid flow of weld pool during electron beam freeform fabrication of Ti-6-Al-4-V alloy. *Int. J. Heat Mass Transf.* **2014**, *78*, 203–215. [CrossRef]
14. Ding, D.; Pan, Z.; Cuiuri, D.; Li, H.; van Duin, S.; Larkin, N. Bead modelling and implementation of adaptive MAT path in wire and arc additive manufacturing. *Robot. Comput. Integr. Manuf.* **2016**, *39*, 32–42. [CrossRef]
15. Greitemeier, D. *Untersuchung der Einflussparameter Auf die Mechanischen Eigenschaften von Additiv Gefertigtem TiAl6V4*; Springer Vieweg: Wiesbaden, Germany, 2016. [CrossRef]
16. Fadida, R.; Rittel, D.; Shirizly, A. Dynamic mechanical behavior of additively manufactured Ti_6Al_4V with controlled voids. *J. Appl. Mech. Trans.* **2015**, *82*. [CrossRef]
17. Kasperovich, G.; Haubrich, J.; Gussone, J.; Requena, G. Correlation between porosity and processing parameters in $TiAl_6V_4$ produced by selective laser melting. *Mater. Des.* **2016**, *10*, 160–170. [CrossRef]
18. Wu, B.; Pan, Z.; Ding, D.; Cuiuri, D.; Li, H.; Xu, J.; Norrish, J. A review of the wire arc additive manufacturing of metals: Properties, defects and quality improvement. *J. Manuf. Process.* **2018**, *35*, 127–139. [CrossRef]
19. DebRoy, T.; Wei, H.L.; Zuback, J.S.; Mukherjee, T.; Elmer, J.W.; Mileweski, J.O.; Beese, A.M.; Wilson-Heid, A.; De, A.; Zhang, W. Additive manufacturing of metallic components—Process, structure and properties. *Prog. Mater. Sci.* **2018**, *92*, 112–224. [CrossRef]
20. Agius, D.; Kourousis, K.; Wallbrink, C. A review of the as-built SLM Ti-6Al-4V mechanical properties towards achieving fatigue resistant designs. *Metals* **2018**, *8*, 75. [CrossRef]
21. Tiferet, E.; Ganor, M.; Zolotaryov, D.; Garkun, A.; Hadjadj, A.; Chonin, M.; Ganor, Y.; Noiman, D.; Halevy, I.; Tevet, O.; et al. Mapping the tray of electron beam melting of Ti-6Al-4V: Properties and microstructure. *Materials* **2019**, *12*, 1470. [CrossRef] [PubMed]
22. Leyens, C.; Peters, M. *Titanium and Titanium Alloys: Fundamentals and Applications*; Wiley-VCG Verlag GmbH & CoKGaA: Weinheim, Germany, 2003. [CrossRef]
23. Lütjering, G.; Williams, J. *Titanium*; Springer: Berlin, Germany, 2007. [CrossRef]
24. Zhang, L.C.; Liu, Y.; Li, S.; Hao, Y. Additive manufacturing of titanium alloys by electron beam melting: A review. *Adv. Eng. Mater.* **2017**, *20*, 1700842. [CrossRef]
25. Roberts, I.A.; Wang, C.J.; Esterlein, R.; Stanford, M.; Mynors, D.J. A three-dimensional finite element analysis of the temperature field during laser melting of metal powders in additive layer manufacturing. *Int. J. Mach. Tools Manuf.* **2009**, *49*, 916–923. [CrossRef]
26. Li, Y.; Gu, D. Parametric analysis of thermal behavior during selective laser melting additive manufacturing of aluminum alloy powder. *Mater. Des.* **2014**, *63*, 856–867. [CrossRef]
27. Murr, L.E.; Esquivel, E.V.; Quinones, S.A.; Gaytan, S.M.; Lopez, M.I.; Martinez, E.Y.; Medina, F.; Hernandez, D.H.; Martinez, E.; Stafford, S.W.; et al. Microstructure and mechanical properties of electron beam-rapid manufactured Ti-6Al-4V biomedical prototypes compared to wrought Ti-6Al-4V. *Mater. Charact.* **2008**, *60*, 96–105. [CrossRef]

28. Tan, X.; Kok, Y.; Tan, Y.J.; Descoins, M.; Mangelinck, D.; Tor, S.B.; Leong, K.F.; Chua, C.K. Graded microstructure and mechanical properties of additive manufactured Ti-6Al-4V via electron beam melting. *Acta Mater.* **2015**, *97*, 1–16. [CrossRef]
29. Zhai, Y.; Galarraga, H.; Lados, D.A. Microstructure, static properties, and fatigue crack growth mechanisms in Ti-6Al-4V fabricated by additive manufacturing: LENS and EBM. *Eng. Fail. Anal.* **2016**, *69*, 3–14. [CrossRef]
30. Fachinotti, V.D.; Cardona, A.; Baufeld, B.; Van der Biest, O. Finite-element modelling of heat transfer in shaped metal deposition and experimental validation. *Acta Mater.* **2012**, *60*, 6621–6630. [CrossRef]
31. Al-Bermani, S.S.; Blackmore, M.L.; Zhang, W.; Todd, I. The origin of microstructural diversity, texture, and mechanical properties in electron beam melted Ti-6Al-4V. *Metall. Mater. Trans.* **2010**, *41*, 3422–3434. [CrossRef]
32. Brandl, E.; Schoberth, A.; Leyens, C. Morphology, microstructure, and hardness of titanium (Ti-6Al-4V) blocks deposited by wire-feed additive layer manufacturing (ALM). *Mater. Sci. Eng. A* **2012**, *532*, 295–307. [CrossRef]
33. Baufeld, B.; Brandl, E.; Van der Biest, O. Wire based additive layer manufacturing: Comparison of microstructure and mechanical properties of Ti-6Al-4V components fabricated by laser-beam deposition and shaped metal deposition. *J. Mater. Process. Technol.* **2011**, *211*, 1146–1158. [CrossRef]
34. Wang, J.; Lin, X.; Li, J.; Hu, Y.; Zhou, Y.; Wang, C.; Li, Q.; Huang, W. Effects of deposition strategies on macro/microstructure and mechanical properties of wire and arc additive manufactured Ti–6Al–4V. *Mater. Sci. Eng. A* **2019**, *754*, 735–749. [CrossRef]
35. Wang, J.; Lin, X.; Wang, J.; Yang, H.; Zhou, Y.; Wang, C.; Li, Q.; Huang, W. Grain morphology evolution and texture characterization of wire and arc additive manufactured Ti-6Al-4V. *J. Alloys Compd.* **2018**, *768*, 97–113. [CrossRef]
36. DVS 2713:2016-04. *Schweißen von Titanwerkstoffen Werkstoffe—Prozesse—Fertigung—Prüfung und Bewertung von Schweißverbindungen*; DVS-Verlag GmbH: Düsseldorf, Germany, 2016.
37. DIN 65084:1990-04. *Luft-und Raumfahrt Wärmebehandlung von Titan und Titan-Knetlegierungen*; CEN: Brussels, Belgium, 1990.
38. DIN EN ISO 6892-1:2017-02. *Metallische Werkstoffe—Zugversuch—Teil 1: Prüfverfahren bei Raumtemperatur*; CEN: Brussels, Belgium, 2017.
39. DIN EN ISO 148-1:2017-05. *Metallische Werkstoffe—Kerbschlagbiegeversuch nach Charpy—Teil 1: Prüfverfahren*; CEN: Brussels, Belgium, 2017.
40. DIN EN ISO 6507-1:2016-08. *Metallische Werkstoffe—Härteprüfung nach Vickers—Teil 1: Prüfverfahren*; CEN: Brussels, Belgium, 2016.
41. Larson, A.C.; von Dreele, R.B. *General Structure Analysis System (GSAS)*; LANL Report No. LAUR 86-748; Los Alamos National Laboratory: Los Alamos, NM, USA, 2000. Available online: http://www.ncnr.nist.gov/xtal/software/gsas.html (accessed on 6 June 2020).
42. Fahrenwaldt, H.J.; Schuler, V. *Praxiswissen Schweißtechnik*; Vieweg+Teubner: Wiesbaden, Germany, 2006.
43. Suryakumar, S.; Karunakaran, K.P.; Bernard, A.; Chandrasekhar, U.; Raghavender, N.; Sharma, D. Weld bead modeling and process optimization in Hybrid Layered Manufacturing. *CAD Comput. Aided Des.* **2011**, *43*, 331–344. [CrossRef]
44. Ding, D.; Pan, Z.; Cuiuri, D.; Li, H. A multi-bead overlapping model for robotic wire and arc additive manufacturing (WAAM). *Robot. Comput. Integr. Manuf.* **2015**, *31*, 101–110. [CrossRef]
45. Ding, D.; Pan, Z.; Cuiuri, D.; Li, H.; Larkin, N. Adaptive path planning for wire-feed additive manufacturing using medial axis transformation. *J. Clean. Prod.* **2016**, *133*, 942–952. [CrossRef]
46. Jin, Y.; He, Y.; Fu, J.; Gan, W.; Lin, Z. Optimization of tool-path generation for material extrusion-based additive manufacturing technology. *Addit. Manuf.* **2014**, *1*, 32–47. [CrossRef]
47. Jin, Y.; He, Y.; Fu, J.; Zhang, A.; Du, J. A non-retraction path planning approach for extrusion-based additive manufacturing. *Robot. Comput. Integr. Manuf.* **2017**, *48*, 132–144. [CrossRef]
48. Ding, D.; Pan, Z.; Cuiuri, D.; Li, H. A practical path planning methodology for wire and arc additive manufacturing of thin-walled structures. *Robot. Comput. Integr. Manuf.* **2015**, *34*, 8–19. [CrossRef]
49. Ding, D.; Shen, C.; Pan, Z.; Cuiuri, D.; Li, H.; Larkin, N.; van Duin, S. Towards an automated robotic arc-welding-based additive manufacturing system from CAD to finished part. *Comput. Des.* **2016**, *73*, 66–75. [CrossRef]

50. Stockinger, J.; Wiednig, C.A.; Enzinger, N.; Sommitsch, C.; Huber, D.; Stockinger, M. Additive Manufacturing via Cold Metal Transfer. In Proceedings of the Metal Additive Manufacturing Conference, Linz, Austria, 24–25 November 2016.
51. Plangger, J.; Schabhüttl, P.; Vuherer, T.; Enzinger, N. CMT additive manufacturing of a high strength steel alloy for application in crane construction. *Metals* **2019**, *9*, 650. [CrossRef]
52. Chen, T.; Pang, S.; Tang, Q.; Suo, H.; Gong, S. Evaporation Ripped Metallurgical Pore in Electron Beam Freeform Fabrication of Ti-6-Al-4-V. *Mater. Manuf. Process.* **2016**, *31*, 1995–2000. [CrossRef]
53. Semiatin, S.L.; Ivanchenko, V.G.; Akhonin, S.V.; Ivasishin, O.M. Diffusion models for evaporation losses during electron-beam melting of alpha/beta-titanium alloys. *Metall. Mater. Trans. B Process. Metall. Mater. Process. Sci.* **2004**, *35*, 235–245. [CrossRef]
54. Juechter, V.; Scharowsky, T.; Singer, R.F.; Körner, C. Processing window and evaporation phenomena for Ti-6Al-4V produced by selective electron beam melting. *Acta Mater.* **2014**, *76*, 252–258. [CrossRef]
55. Zhang, G.; Chen, J.; Zheng, M.; Yan, Z.; Lu, X.; Lin, X.; Huang, W. Element vaporization of Ti-6AL-4V alloy during selective laser melting. *Metals* **2020**, *10*, 435. [CrossRef]
56. Sequeira Almeida, P.M.; Williams, S. Innovative process model of Ti-6Al-4V additive layer manufacturing using cold metal transfer (CMT). In Proceedings of the 21st Annual International Solid Freeform Fabrication Symposium, Austin, TX, USA, 9–11 August 2010.
57. Bermingham, M.J.; StJohn, D.H.; Krynen, J.; Tedman-Jones, S.; Dargusch, M.S. Promoting the columnar to equiaxed transition and grain refinement of titanium alloys during additive manufacturing. *Acta Mater.* **2019**, *168*, 261–274. [CrossRef]
58. Åkerfeldt, P.; Antti, M.-L.; Pederson, R. Influence of microstructure on mechanical properties of laser metal wire-deposited Ti-6Al-4V. *Mater. Sci. Eng. A* **2016**, *674*, 428–437. [CrossRef]
59. Ho, A.; Zhao, H.; Fellowes, J.W.; Martina, F.; Davis, A.E.; Prangnell, P.B. On the origin of microstructural banding in Ti-6Al4V wire-arc based high deposition rate additive manufacturing. *Acta Mater.* **2019**, *166*, 306–323. [CrossRef]
60. Haubrich, J.; Gussone, J.; Barriobero-Vila, P.; Kürnsteiner, P.; Jägle, E.A.; Raabe, D.; Schell, N.; Requena, G. The role of lattice defects, element partitioning and intrinsic heat effects on the microstructure in selective laser melted Ti-6Al-4V. *Acta Mater.* **2019**, *167*, 136–148. [CrossRef]
61. Daymond, M.R.; Bourke, M.A.M.; von Dreele, R.B. Use of Rietveld refinement to fit a hexagonal crystal structure in the presence of elastic and plastic anisotropy. *J. Appl. Phys.* **1999**, *85*, 739–747. [CrossRef]
62. Lundbäck, A.; Pederson, R.; Colliander, M.H.; Brice, C.; Steuwer, A.; Heralic, A.; Buslaps, T.; Lindgren, L.-E. Modeling and experimental measurement with synchrotron radiation of residual stresses in laser metal deposited Ti-6Al-4V. In *Proceedings of the 13th World Conference on Titanium, San Diego, CA, USA, 16–20 August 2015*; Wiley: Hoboken, NJ, USA, 2016. [CrossRef]
63. ASTM F136-13. *Standard Specification for Wrought Titanium-6Aluminum-4Vanadium ELI (Extra Low Interstitial) Alloy for Surgical Implant Applications (UNS R56401)*; ASTM International: West Conshohocken, PA, USA, 2013.
64. Grell, W.A.; Solis-Ramos, E.; Clark, E.; Lucon, E.; Garboczi, E.J.; Predecki, P.K.; Loftus, Z.; Kumosa, M. Effect of powder oxidation on the impact toughness of electron beam melting Ti-6Al-4V. *Addit. Manuf.* **2017**, *17*, 123–134. [CrossRef]
65. Yasa, E.; Deckers, J.; Kruth, J.-P.; Rombouts, M.; Luyten, J. Charpy impact testing of metallic selective laser melting parts. *Virtual Phys. Prototyp.* **2010**, *5*, 89–98. [CrossRef]
66. Wu, M.W.; Lai, P.H. The positive effect of hot isostatic pressing on improving the anisotropies of bending and impact properties in selective laser melted Ti-6Al-4V alloy. *Mater. Sci. Eng. A* **2016**, *658*, 429–438. [CrossRef]
67. Sliva, A.P.; Dragunov, V.K.; Terentyev, E.V.; Goncharov, A.L. EBW of aluminium alloys with application of electron beam oscillation. *Proc. J. Phys. Conf. Ser.* **2018**, *1089*, 012005. [CrossRef]
68. Gudenko, A.V.; Sliva, A.P. Influence of electron beam oscillation parameters on the formation of details by electron beam metal wire deposition method. *Proc. J. Phys. Conf. Ser.* **2018**, *1109*, 012037. [CrossRef]
69. Dragunov, V.K.; Goryachkina, M.V.; Gudenko, A.V.; Sliva, A.P.; Shcherbakov, A.V. Investigation of the optimal modes of electron-beam wire deposition. In Proceedings of the IOP Conference Series: Materials Science and Engineering, Tomsk, Russia, 3–7 September 2019. [CrossRef]
70. Zhao, J.; Zhang, B.; Li, X.; Li, R. Effects of metal-vapor jet force on the physical behavior of melting wire transfer in electron beam additive manufacturing. *J. Mater. Process. Technol.* **2015**, *220*, 243–250. [CrossRef]

71. Hu, R.; Luo, M.; Liu, T.; Liang, L.; Huang, A.; Trushnikov, D.; Karunakaran, K.P.; Pang, S. Thermal fluid dynamics of liquid bridge transfer in laser wire deposition 3D printing. *Sci. Technol. Weld. Join.* **2018**, *24*, 401–411. [CrossRef]
72. Xiong, J.; Zhang, G.; Gao, H.; Wu, L. Modeling of bead section profile and overlapping beads with experimental validation for robotic GMAW-based rapid manufacturing. *Robot. Comput. Integr. Manuf.* **2013**, *29*, 417–423. [CrossRef]
73. Dong, Z.; Liu, Y.; Wen, W.; Ge, J.; Liang, J. Effect of hatch spacing on melt pool and as-built quality during selective laser melting of stainless steel: Modeling and experimental approaches. *Materials* **2019**, *12*, 50. [CrossRef]
74. Wang, P.; Nai, M.L.S.; Sin, W.J.; Lu, S.; Zhang, B.; Bai, J.; Song, J.; Wei, J. Effect of overlap distance on the microstructure and mechanical properties of in situ welded parts built by electron beam melting process. *J. Alloys Compd.* **2019**, *772*, 247–255. [CrossRef]
75. Cao, Y.; Zhu, S.; Liang, X.; Wang, W. Overlapping model of beads and curve fitting of bead section for rapid manufacturing by robotic MAG welding process. *Robot. Comput. Integr. Manuf.* **2011**, *27*, 641–645. [CrossRef]
76. Li, Y.; Han, Q.; Zhang, G.; Horváth, I. A layers-overlapping strategy for robotic wire and arc additive manufacturing of multi-layer multi-bead components with homogeneous layers. *Int. J. Adv. Manuf. Technol.* **2018**, *96*, 3331–3344. [CrossRef]
77. Nguyen, L.; Buhl, J.; Bambach, M. Multi-bead overlapping models for tool path generation in wire-arc additive manufacturing processes. *Procedia Manuf.* **2020**, *47*, 1123–1128. [CrossRef]
78. Li, Y.; Sun, Y.; Han, Q.; Zhang, G.; Horváth, I. Enhanced beads overlapping model for wire and arc additive manufacturing of multi-layer multi-bead metallic parts. *J. Mater. Process. Technol.* **2018**, *252*, 838–848. [CrossRef]
79. Aiyiti, W.; Zhao, W.; Lu, B.; Tang, Y. Investigation of the overlapping parameters of MPAW-based rapid prototyping. *Rapid Prototyp. J.* **2006**, *12*, 165–172. [CrossRef]
80. Nikam, S.H.; Jain, N.K. Finite element simulation of pre-heating effect on melt pool size during micro-plasma transferred arc deposition process. *IOP Conf. Ser. Mater. Sci. Eng.* **2018**, *389*, 012006. [CrossRef]
81. Liu, S.; Shin, Y.C. Additive manufacturing of Ti6Al4V alloy: A review. *Mater. Des.* **2019**, *164*, 107552. [CrossRef]
82. Edwards, P.; O'Conner, A.; Ramulu, M. Electron beam additive manufacturing of titanium components: Properties and performance. *J. Manuf. Sci. Eng.* **2013**, *135*. [CrossRef]
83. Chong, Y.; Bhattacharjee, T.; Yi, J.; Shibata, A.; Tsuji, N. Mechanical properties of fully martensite microstructure in Ti-6Al-4V alloy transformed from refined beta grains obtained by rapid heat treatment (RHT). *Scr. Mater.* **2017**, *138*, 66–70. [CrossRef]

© 2020 by the authors. Licensee MDPI, Basel, Switzerland. This article is an open access article distributed under the terms and conditions of the Creative Commons Attribution (CC BY) license (http://creativecommons.org/licenses/by/4.0/).

Article

Changing in Larch Sapwood Extractives Due to Distinct Ionizing Radiation Sources

Thomas Schnabel [1,*], Marius Cătălin Barbu [1,2], Eugenia Mariana Tudor [1,2] and Alexander Petutschnigg [1,3,4]

- [1] Department Forest Products Technology and Timber Constructions, Salzburg University of Applied Sciences, Markt 136a, 5431 Kuchl, Austria; marius.barbu@fh-salzburg.ac.at (M.C.B.); Eugenia.tudor@fh-salzburg.ac.at (E.M.T.); alexander.petutschnigg@fh-salzburg.ac.at (A.P.)
- [2] Faculty of Furniture Design and Wood Engineering, Transilvania University of Brasov, B-dul. Eroilor nr. 29, 500036 Brasov, Romania
- [3] Salzburg Center for Smart Materials, c/o Department of Chemistry and Physics of Materials, Paris Lodron University of Salzburg, Jakob-Harringer-Strasse 2A, 5020 Salzburg, Austria
- [4] Department of Material Sciences and Process Engineering, University of Natural Resources and Life Sciences (BOKU), Konrad Lorenz-Straße 24, 3340 Tulln, Austria
- * Correspondence: thomas.schnabel@fh-salzburg.ac.at

Citation: Schnabel, T.; Barbu, M.C.; Tudor, E.M.; Petutschnigg, A. Changing in Larch Sapwood Extractives Due to Distinct Ionizing Radiation Sources. *Materials* 2021, 14, 1613. https://doi.org/10.3390/ma14071613

Academic Editor: Katia Vutova

Received: 19 February 2021
Accepted: 22 March 2021
Published: 26 March 2021

Publisher's Note: MDPI stays neutral with regard to jurisdictional claims in published maps and institutional affiliations.

Copyright: © 2021 by the authors. Licensee MDPI, Basel, Switzerland. This article is an open access article distributed under the terms and conditions of the Creative Commons Attribution (CC BY) license (https://creativecommons.org/licenses/by/4.0/).

Abstract: Wood extractives have an influence on different material properties. This study deals with the changes in wood extractives of larch sapwood due to two different low doses of energy irradiations. Electron beam irradiation (EBI) and γ-ray irradiation treatments were done by using two industrial processes. After the different modifications the extractions were performed with an accelerated solvent extractor (ASE) using hexane and acetone/water. The qualitative and quantitative chemical differences of irradiated larch sapwood samples were analysed using gas chromatography–mass spectrometry (GC-MS) and Fourier-transform infrared spectroscopy (FT-IR) vibrational spectroscopy methods. The yields of the quantitative extractions decreased due to the two different irradiation processes. While the compounds extracted with nonpolar solvent from wood were reduced, the number of compounds with polar functionalities increased based on the oxidation process. Quantitatively, resin acids and polyphenols were highly affected when exposed to the two irradiation sources, leading to significant changes (up, down) in their relative amount. Furthermore, two new substances were found in the extracts of larch sapwood samples after EBI or γ-ray treatments. New insight into the different effects of larch sapwood and wood extractives by EBI and γ-ray was gained in this study.

Keywords: EBI; γ-ray; GC-MS; FT-IR; larch sapwood; wood extractives

1. Introduction

The effects of irradiation of different wood polymers have been studied since the early 1960s [1]. The irradiated polymers undergo various complex changes in chemistry [2–5]. By varying the irradiation doses and radiation sources, the material properties can be differently influenced, and therefore can be used for different applications connected to the functionalisation of cellulose and lignin [6–8]. The wood extractives may influence these chemical changes caused by irradiation. This study deals with the changes in wood extractives of larch sapwood due to low doses of irradiation.

In general, two different reasons for irradiation treatments can be distinguished. On the one hand, ionising radiation was investigated for a pre-treatment pulping process to separate the cellulose and lignin with a low accumulated dose and less chemicals compared to conventional processes [5,9]. A few efforts have been made to force the degradation of lignin without a cross-linking process [2].

On the other hand, the irradiation has been used to functionalise different properties for the enhancement of wood [6,10]. After a low dose of electron beam irradiation of wood, the weathering effects were reduced and the colour stabilisation was improved [7]. Condensation reactions occur within the lignin at low radiation dosages and stabilise the wood matrix against different processes [2,6]. Moreover, new interaction between chemical bonds can be generated in cellulose by using ionising radiation [8] and may lead to enhanced properties of wood and cellulose [11]. Through the low absorbed dose of the irradiation, the hemicelluloses (polyoses) are altered and can change the material resistance against compression [6]. Nevertheless, the influence of radiation on wood extractives has not been clarified in detail. Lawniczak and Raczkowski [12] concluded that the extractive prevents the wood for mechanical property losses caused by a radiation process.

In general, the amount of wood extractives is between 1–7% of oven dry (o.d.) weight for European softwood wood species, which is very low compared to the amount of cellulose, hemicellulose, and lignin [13]. However, wood extractives influence different wood properties (e.g., colour and durability) [14] and contain different groups of compounds (e.g., polyphenols and sugars) [15]. Overall, it could be concluded that these various groups may be affected differently by the ionising radiation based on their chemical structure.

In this study, the qualitative and quantitative chemical differences of low doses of EBI and γ-ray on larch sapwood samples were analysed using a gas chromatography–mass spectrometry (GC-MS) method. New insight into the effects of two ionising radiation sources on wood performances will help to understand the significance of the extractives. Furthermore, FT-IR vibrational spectroscopy methods were applied for the ratio assessment of lignin to cellulose within the samples to characterise the possible changes within the two wood components.

2. Materials and Methods

2.1. Material

Seventy-five samples were obtained from larch sapwood (*Larix decidua* (Mill)), which were cut to dimensions of $1.0 \times 1.0 \times 1.0$ cm^3. The samples were divided into three groups of 25 pieces for the various treatments and chemical characterisation after the different irradiations. Before testing, the samples were ground with a cutting mill (Retsch, Haan, Germany) using solid carbon dioxide to pass a mesh of 500 μm. The larch wood powder was then freeze-dried to prevent chemical changes in material during drying with higher temperature.

2.2. Radiations

The EBI treatment was performed under normal air environmental conditions at a dose of 10 kGy with a Rhodotron TT-100 10 MeV electron accelerator (IBA International, Louvain-La-Neuve, Belgium, radiation energy of 10 MeV). The γ-ray irradiation was done in a Gammatron 1500 gamma irradiator (Mediscan, Seibersdorf, Austria) at a dose of 10 kGy by using the decay of ^{60}Co.

2.3. Solid-Liquid Extraction

The extraction of the wood was done with an accelerated solvent extractor (ASE, Dionex Corporation, Sunnycale, CA, USA.) according to Willför et al. [16]. The larch sapwood samples were first extracted with hexane (solvent temperature 90 °C, pressure 13.8 Mpa, 2 × 5 min static cycles) and then with acetone and water in a ratio 95/5 mixture (100 °C, 13.8 Mpa, 2 × 5 min).

2.4. GC-MS Characterisation

Before GC-MS characterisation, different extractives were evaporated using nitrogen gas and silylated to enhance volatility. This GC-MS method was developed according to Wagner et al. [17]. For silylation, the evaporated extractives were first dried in a vacuum oven at 40 °C, and the silylated solvent (80 μL bis(trimethylsilyl)-trifluoracetamide, 20 μL pyridine and 20 μL trimethylsilyl-chloride) were added. After that, the samples were incubated at 70 °C for 45 min. Measurements were performed using a Perkin Elmer

Auto-System CL gas chromatograph (PerkinElmer Inc., Waltham, MA, USA) and an MS device. The GC was equipped with an HP-5 column (length: 25 cm; ID: 0.20 mm; film thickness 0.11 µm) and a flame ionization detector (FID, Agilent Technologies Inc., Santa Clara, CA, USA). The carrying gas was nitrogen at a flow rate of 0.8 mL/min. Moreover, other conditions were: internal oven 120 °C with an increasing rate at 6 °C/min to 320 °C (15 min hold); a split injection with a ratio of 25:1 and a temperature of 250 °C; the detector temperature 310 °C and injection value of 1 µL. The data were investigated based on the mass spectra library created at the Laboratory of Natural Materials Technology at Åbo Akademic University in Turku.

2.5. FT-IR Spectroscopy Analysis

The obtained samples from sapwood of larch were characterised by infrared spectroscopy with a Perkin Elmer FT-IR spectrometer (Waltham, MA, USA) equipped with a Miracle diamond ATR accessory with a 1.8 mm round crystal surface. The spectra were recorded in the wavenumber range between 600 cm^{-1} and 4000 cm^{-1} with 32 scans at a resolution of 4 cm^{-1}. All spectra were ATR and baseline corrected in the wavenumber range between 600 cm^{-1} and 4000 cm^{-1} as well as normalised the absorbance values between 0 and 1.

3. Results and Discussion

The FT-IR spectra from larch sapwood samples without and after different irradiation processes are presented in Figure 1. The four marked band areas show an increase in absorbance resulting from the EBI and γ-ray treated samples. The peaks obtained at 3340 cm^{-1} are an indication of intramolecular hydrogen bond of cellulose [18]. The IR signal in the range between 2933 cm^{-1} and 2835 cm^{-1} corresponded with the asymmetric CH$_2$ valence vibration and the CH$_2$, CH$_2$OH groups in the wood material. Furthermore, the absorbance of the carbonyl groups at 1730 cm^{-1} wavenumbers changes due to irradiation.

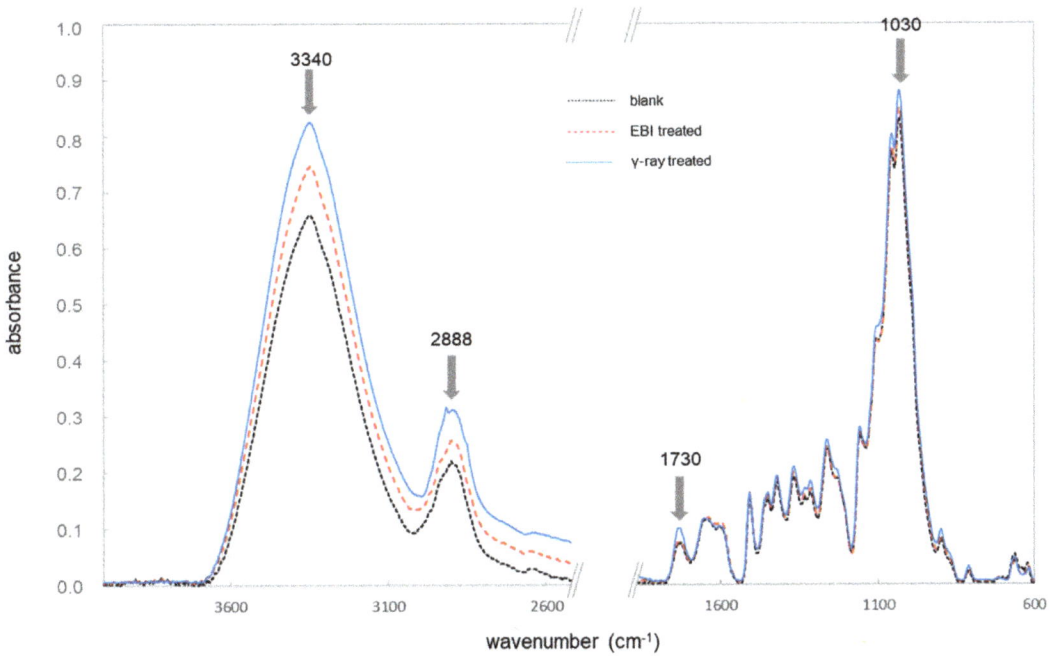

Figure 1. Fourier-transform infrared spectroscopy (FT-IR) spectra of larch sapwood samples with different treatments.

The band at 1030 cm^{-1} corresponds to C-O valence vibration and C$_{alky}$-O ether vibrations of different wood compounds. All these observations show that there is the separation of new molecule types from radical formation, ring, and chain breaking of the wood, mainly from the carbohydrates, which has amount around 67.1% in dry larch wood [13] and has the most important impact on the IR signals. Su et al. [19] mentioned the cellulose and lignin contents of gamma-irradiated bamboo were stable up to 300 kGy, whereas polyoses and extractives were not analysed and could protect the cellulose and lignin from degradation. This behaviour seemed reasonable, as Yang et al. [20] concluded that the cellulose from paper grade bamboo pulp were easily degraded by a ^{60}Co γ-ray radiation. According to Schwanninger et al. [18], the bands around 1510 and 898 cm^{-1} correspond to the lignin and cellulose polymers, respectively. This relative ratio of these IR bands was used to assess the relative lignin content in the wood and its change [7]. Based on this approach, the results of the ratio are 1.88, 1.83, and 1.66 for untreated, EBI, and γ-ray-treated larch wood powder, respectively. While the blank and EBI-treated samples show a similar ratio, the γ-ray-irradiated sample was different and depicts a lower amount of lignin content compared to the others. Based on detailed analysis results that the absorbance of the IR band at around 898 cm^{-1} was higher than the both other sample groups, it can be assumed that the functional group of the carbohydrates increased due to the γ-ray treatment more than the absorbance of the aromatic skeletal vibration at 1510 cm^{-1}.

The extracts were analysed quantitatively after the different treatments in the first steps. The extraction yields of o.d. larch sapwood with hexane differed in a range from 2.27 mg/g of blank to 1.36 mg/g of EBI treated and 1.67 mg/g of γ-ray treated samples, respectively. Through the ionising radiations, the amount of hexane extractable compounds decreased significantly, at which the EBI had the significant effect of the extracts compared to the γ-ray process. Overall, the resin acids appeared to be very sensitive to the irradiation treatments and exhibited the strongest decrease compared to the other og compound studied (Table 1). Based on the detailed analysis of single substances of the blank extracts, EBI and γ-ray-treated samples showed that almost all amounts of single compounds decreased, except palmitic acid, for which values increased for the extract from γ-ray-treated samples. On the one hand, the extractives could be cross-linked to higher molecule mass and/or other wood components resulting from the insolubility of the hexane solvent. On the other hand, the effects of the breaking of chemical bonds of wood extractives may change the compound solubility [21]. The formation of free molecules and broken glucose rings indicate the damage of cellulose. These chemical intermediates can react with the oxygen in the air and result in the change of polarity of components.

Table 1. Main component groups in hexane extracts of different treated larch sapwood samples by gas chromatography–mass spectrometry (GC-MS).

Component Groups	Blank Larch Sapwood (mg/g)	EBI-Treated Larch Sapwood (mg/g)	X-ray-Treated Larch Sapwood (mg/g)
Terpenoid	0.185	0.038	0.104
Alcohol	0.009	0.001	0.003
Fatty acids	0.486	0.124	0.217
Stilbenoid	0.005	0.000	0.000
Resin acids	2.083	0.177	0.547
Aliphatic compounds	0.005	0.001	0.002
Hydroxy resin acids	0.051	0.010	0.023
Lignans	0.062	0.002	0.011
Precursor of Lignin	0.020	0.005	0.010
Unknown	0.962	0.130	0.306

This process could be also seen by using other solvents that are more polar than hexane. Therefore, acetone was used as second solvent to recover polar compounds from the larch sapwood after different treatments. The extraction yields of o.d. larch sapwood with hexane differed in a range from 9.73 mg/g of blank to 4.48 mg/g of EBI treated and 4.80 mg/g of γ-ray treated samples, respectively. Generally, the amount of extract with acetone is higher comparted to the hexane solvent. Nevertheless, the ratio between the different treatments is quite uniform, the EBI- and γ-ray-treated samples showed less extraction yield than the blank samples. Almost all amounts of component groups increased due to the different treatment without the exception of the polyphenol groups (Table 2)—that extraction yield decreased very strongly.

Table 2. Main component groups in acetone extracts of different treated larch sapwood samples by GC-MS.

Component Groups	Blank Larch Sapwood (mg/g)	EBI-Treated Larch Sapwood (mg/g)	X-ray-Treated Larch Sapwood (mg/g)
Carboxylic acids	0.051	0.155	0.348
Phenylpropanoid	0.242	0.253	0.654
Polyhydric alcohols	0.027	0.195	0.486
Single sugars	0.415	0.832	2.043
Aliphatic acids	0.064	0.096	0.323
Resin acids	0.028	0.061	0.225
Polyphenols	7.019	0.499	2.659
Phytosterine	0.028	0.005	0.019

Especially, dihydrokaempferol, (+)-catechin, and taxifolin from the flavonoid group were mostly affected due to the irradiation treatments. These compounds are known as antioxidants for inhibiting free radicals based on different activities (e.g., hydroxyl substituents, peroxyl reaction) in materials [22]. In the extracts of the irradiated samples, a modified taxifolin structure was quantitatively identified, which was not found in the blank samples and supports the idea of radical scavenging of flavonoids and further cross-linking effects. Also, Błaszak et al. [23] determined a high degradation rate of the flavan-3-ol substance of (+)-catechin due to 10 kGy EB irradiation compared to the gallic acid (phenolic acid), which was not affected by the EBI treatment up to this absorbed dose. The hydroxybenzoic acids like e.g., vanillic acid are possible metabolites of the phenylpropanoid pathway, and the extraction amount of group of phenylpropanoid was stable for the blank samples and the EBI-treated samples and increased strongly for the γ-ray-treated samples. The acetone extracted compounds with one or two guaiacyl units in its molecular structure were significantly increased in quantity, whereas the component amount of benzene-1,2-dicarbocylic acid with two aliphatic rest in the structure decreased from 0.164 to 0.038 mg/g by the two irradiation treatments. It can be accepted that the electrons or γ-ray starts the degradation of the aliphatic structure and some of these damaged compounds arranged to guaiacyl derivates with one or two phenyl groups.

Compared to the polyphenol groups, the amount of polyhydric alcohols increased due to the EBI and even more for the γ-ray irradiation. Cyclitols like pinitol (3-O-methyl-chiro-inositol) and myo-inositol (1,2,3,4,5,6-Hexahydroxycyclohexane) were found in a larger amount after irradiations than before it. Also, further sugar alcohols were determined after the ionising radiation. All these various compounds show a carbohydrate structure, and it can be assumed that these products were generated from the cellulose and polyoses by the irradiation processes.

Furthermore, two new substances were found in the extracts of larch sapwood samples after EBI or γ-ray treatments. The lignan α-conidendric acid was characterised by the extracts from both treatments, and isohydroxymatairesinol (iso-HMR) was identified only in the extract from the γ-ray treated samples. Both of these molecules have two phenyl rings within the structure that can be arranged from other polyphenols or lignans, which was shown from the decreased in polyphenols contents in the extracts. Zule et al. [24] mentioned the secoisolariciresinol was the dominantly substances of lignan in larch heartwood; however, larch sapwood was not investigated in the same study. Nevertheless, Nisula [25] concluded that sapwood contained only traces of lignan compared to the concentration in larch heartwood without giving some values. Also, the results in this study indicate that secisolariciresinol is not present in high amounts compared to the other polyphenols, such as dihydrokaempferol or taxifolin, at least in these materials which were characterised. However, the lignans were detectable and show the effects of the irradiation treatments.

4. Conclusions

Various chemical changes in larch sapwood extractives due to different ionization irradiation of electron beam and γ-ray technologies were observed by a FT-IR vibrational spectroscopy and GC-MS methods. The extraction yields of the samples were significantly reduced after the different treatments comparted to the unirradiated ones.

This lower amount of leaching out of wood extractives of irradiated samples could stabilise the material properties against different influences, such as weathering effects or microorganism attack. The quantitative most important substances of resin acids and polyphenols groups were highly affected of the two irradiation processes. Furthermore, two new substances were found in the extracts of larch sapwood samples after EBI or γ-ray treatments.

These results demonstrate the importance of the wood extractives on wood treatment by two ionising irradiation processes, which might have an impact on material performance.

The findings show that the different ionising radiation sources could be used as an interesting technique for changing the chemical compounds in wood.

Author Contributions: Conceptualization, T.S. and M.C.B.; methodology, T.S. and E.M.T.; software, T.S.; investigation, T.S., M.C.B. and E.M.T.; resources, A.P.; writing—original draft preparation, T.S.; writing—review and editing, M.C.B., E.M.T., and A.P.; project administration, T.S.; funding acquisition, A.P. All authors have read and agreed to the published version of the manuscript.

Funding: This research was co-funded by the European Funds for Regional Development (EFRE), Austria Wirtschafts Service (AWS), and the region of Salzburg for the support in the development of the Salzburg Center for Smart Materials. Furthermore, this work was supported from the federal state of Salzburg under the grant 'Holz.Aktiv'.

Institutional Review Board Statement: Not applicable.

Informed Consent Statement: Not applicable.

Acknowledgments: The authors thank Kerstin Wagner, MSc for doing the GC-MS measurements in her Short-Term Scientific Mission (STSM) of the COST Action FP 1407. We would like to thank Stefan Willför from the Åbo Akademi University for kindly providing his facilities for the identification of extractives.

Conflicts of Interest: The authors declare no conflict of interest. The funders had no role in the design of the study; in the collection, analyses, or interpretation of data; in the writing of the manuscript, or in the decision to publish the results.

References

1. Saeman, J.F.; Millett, M.A.; Lawton, E.J. Effect of high-energy cathode rays on cellulose. *Ind. Eng. Chem.* **1952**, *44*, 2848–2852. [CrossRef]
2. Fischer, K.; Goldberg, W. Changes in lignin and cellulose by irradiation. *Makromol. Chem. Marcomol. Symp.* **1987**, *12*, 303–322. [CrossRef]
3. Buremester, A. The improvement of wood by radiation-initiated polymerisation of monomer plastic. *Holz Roh Werkst.* **1967**, *25*, 11–25.
4. Seifert, K. Zur Chemie gammabestrahlten Holzes. *Holz Roh Werkst.* **1964**, *22*, 267–275. [CrossRef]
5. Hoffmann, P.; Schweers, W. On the hydrogenolysis of lignin. 10. Comparative hydrogenolyses of lignins, lignosulfonic acid, and lignosulfonate model compounds under irradiation with γ-rays. *Paperi Ja Puu.* **1976**, *58*, 227–244.
6. Schnabel, T.; Huber, H.; Grünewald, T.; Lichtenegger, H.C.; Petutschnigg, A. Changes in mechanical and chemical wood properties by electron beam irradiation. *Appl. Surf. Sci.* **2015**, *332*, 704–709. [CrossRef]

7. Schnabel, T.; Huber, H. Improving the weathering on larch wood samples by electron beam irradiation (EBI). *Holzforschung* **2014**, *68*, 679–683. [CrossRef]
8. Baccaro, S.; Carewska, M.; Casieri, C.; Cemmi, A.; Lepore, A. Structure modifications and interaction with moisture in γ-irradiated pure cellulose by thermal analysis and infrared spectroscopy. *Polym. Degrad. Stabil.* **2013**, *98*, 2005–2010. [CrossRef]
9. LaVerne, J.A.; Driscoll, M.S.; Al-Sheikhly, M. Radiation stability of lignocellulosic material compounds. *Radiat. Phys. Chem.* **2000**, *171*, 108716. [CrossRef]
10. Huber, H.; Haas, R.; Petutschnigg, A.; Grüll, G.; Schnabel, T. Changes in wettability of wood surface using electron beam irradiation. *Wood Mater. Sci. Eng.* **2020**, *15*, 237–240. [CrossRef]
11. Henniges, U.; Hasani, M.; Potthast, A.; Westman, G.; Rosenau, T. Electron beam irradiation of cellulosic materials—Opportunities and limitations. *Materials* **2013**, *6*, 1584–1598. [CrossRef]
12. Lawniczak, M.; Razkowski, J. The influence of extractives on the radiation stability of wood. *Wood Sci. Technol.* **1970**, *4*, 45–49. [CrossRef]
13. Fengel, D.; Grosser, D. Chemical composition of softwoods and hardwoods—A bibliographical review. *Holz Roh Werkst.* **1975**, *33*, 32–34. [CrossRef]
14. Fengel, D.; Wegner, G. *Wood Chemistry Ultrastructure Reactions*; Verlag Kessel: Remagen, Germany, 2003.
15. Wagner, K.; Roth, C.; Willför, S.; Musso, M.; Petutschigg, A.; Oostingh, G.J.; Schnabel, T. Identification of antimicrobial compounds in different hydrophilic larch bark extracts. *BioRescources* **2019**, *14*, 5807–5815.
16. Willför, S.M.; Hemming, J.; Raunanen, M.; Holmbom, B. Phenolic and lipophilic extractives in Scots pine knots and stemwood. *Holzforschung* **2003**, *57*, 359–372. [CrossRef]
17. Wagner, K.; Musso, M.; Kain, S.; Willför, S.; Petutschigg, A.; Schnabel, T. Larch wood residues valorization through extraction and utilization of high value-added products. *Polymers* **2020**, *12*, 359. [CrossRef]
18. Schwanninger, M.; Rodrigues, J.C.; Pereira, H.; Hinterstoisser, B. Effects of short-time vibratory ball milling on the shape of FT-IR of wood and cellulose. *Vib. Spectrosc.* **2004**, *36*, 23–40. [CrossRef]
19. Su, F.; Jiang, J.; Sun, O.; Lu, F. Changes in chemical composition and microstructure of bamboo after gamma ray irradiation. *BioRescources* **2014**, *9*, 5794–5800.
20. Yang, G.; Zhang, Y.; Wei, M.; Shao, H.; Hu, X. Influence of γ-ray radiation on the structure and properties of paper grade bamboo pulp. *Carbohydr. Polym.* **2010**, *81*, 114–119. [CrossRef]
21. Polvi, J.; Nordlund, K. Low-energy irradiation effects in cellulose. *J. Appl. Phys.* **2014**, *115*, 023521. [CrossRef]
22. Heim, K.E.; Tagilferr, A.R.; Bobilya, D.J. Flavonoid antioxidants: Chemistry, metabolism and structure-activity relationships. *J. Nutr. Biochem.* **2002**, *13*, 572–584. [CrossRef]
23. Błaszak, M.; Nowak, A.; Lachowicz, S.; Migdał, W.; Ochmian, I. E-Beam irradiation and ozonation as an alternative to the sulphuric method of wine preservation. *Molecules* **2019**, *24*, 3406. [CrossRef] [PubMed]
24. Zule, J.; Čufar, K.; Tišler, V. Hydrophilic extractives in heartwood of European larch (*Larix decidua* Mill). *Drv. Ind.* **2016**, *67*, 363–370. [CrossRef]
25. Nisula, L. Wood Extractives in Conifers. A Study of Steamwood and Knots of Industrially Important Species. Ph.D. Thesis, Åbo Akademi University, Turku, Finland, 2018.

Article

Thermal Behavior of Ti-64 Primary Material in Electron Beam Melting Process

Jean-Pierre Bellot *, Julien Jourdan, Jean-Sébastien Kroll-Rabotin, Thibault Quatravaux and Alain Jardy

Institut Jean Lamour—UMR CNRS 7198, LabEx DAMAS, Campus Artem, Université de Lorraine, 2 allée André Guinier, 54000 Nancy, France; julien.jourdan@univ-lorraine.fr (J.J.); jean-sebastien.kroll-rabotin@univ-lorraine.fr (J.-S.K.-R.); thibault.quatravaux@univ-lorraine.fr (T.Q.); alain.jardy@univ-lorraine.fr (A.J.)
* Correspondence: jean-pierre.bellot@univ-lorraine.fr; Tel.: +33-372-744-917

Abstract: The Electron Beam Melting (EBM) process has emerged as either an alternative or a complement to vacuum arc remelting of titanium alloys, since it is capable of enhancing the removal of exogenous inclusions by dissolution or sedimentation. The melting of the primary material is a first step of this continuous process, which has not been studied so far and is investigated experimentally and numerically in the present study. Experiments have been set up in a 100 kW laboratory furnace with the aim of analyzing the effect of melting rate on surface temperature of Ti-64 bars. It was found that melting rate is nearly proportional to the EB power while the overheating temperature remains roughly independent of the melting rate and equal to about 100 °C. The emissivity of molten Ti-64 was found to be 0.22 at an average temperature of about 1760 °C at the tip of the bar. In parallel, a mathematical model of the thermal behavior of the material during melting has been developed. The simulations revealed valuable results about the melting rate, global heat balance and thermal gradient throughout the bar, which agreed with the experimental values to a good extent. The modeling confirms that the overheating temperature of the tip of the material is nearly independent of the melting rate.

Keywords: melting; electron beam; melting temperature; numerical simulation

1. Introduction

Among the secondary remelting techniques, the EBM (Electron Beam Melting) process applies a high-power electron beam on the metallic material for melting, refining and controlling the casting and solidification stages. These operations ensure both the purification of the metal as it is gradually melted and the controlled solidification of the ingot in terms of structure and chemical homogeneity.

The EBM process can be used in two ways: Electron Beam Cold-Hearth Melting (EBCHM) and Electron Beam Drip Melting (EBDM). Schematic diagrams of these techniques are shown in Figure 1.

The main difference between these two methods is the inclusion of a cold hearth (circled in red in Figure 1a) in the EBCHM process as an intermediate refining stage between the melting and solidification steps. In the melting step, the electron beam heats and melts the bar tip. In the final solidification step, the molten metal is cast into a withdrawing water-cooled copper mold and solidified to form an ingot. All these operations are performed inside a vacuum chamber (10^{-4} to 10^{-3} mbar) in order to guarantee proper operation of the electron guns and to avoid alloy contamination. The EBM methods have a number of advantages when compared to the classic vacuum arc remelting (VAR) process, that lead to a better ingot quality. In particular, melting is conducted in a higher vacuum and for a longer time, thus enabling more complete degassing and dissolution of exogenous inclusions such as low-density inclusions known as hard-alpha [1,2].

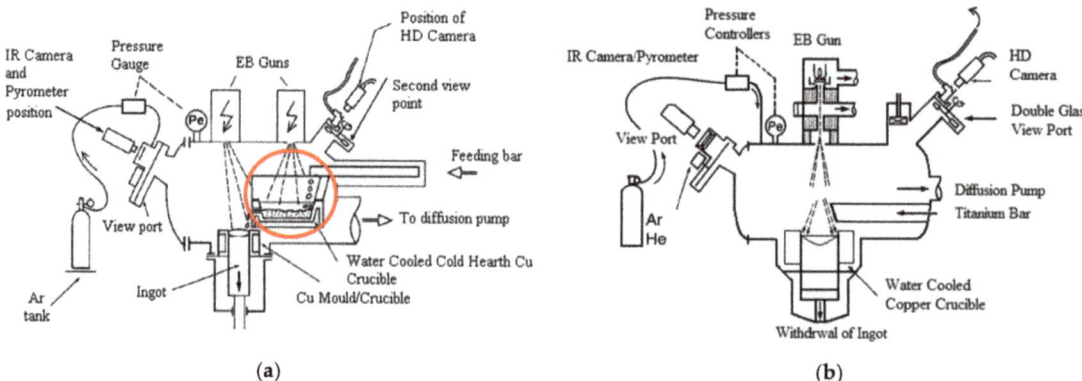

Figure 1. Schematic diagrams of: (**a**) EBCHM; (**b**) EBDM.

The EBM processes are commonly used for manufacturing refractory or reactive metals, such as niobium and titanium [2,3]. Since these techniques essentially concern very high-performance materials used in leading-edge applications demanding exceptional reliability, it is extremely important to choose the most appropriate operating parameters in order to achieve optimum product quality. Despite experimental and numerical works having been performed on the refining and solidification steps [2–5], the thermal behavior of the raw material during the melting stage has only been investigated in the case of alternative processes such as vacuum arc remelting in the literature [6,7] although it plays an important role on the process operation. The surface temperature of the tip of the load activates the volatilization mechanism, which controls the material refining and the losses in alloying elements such as aluminum in Ti-64. Furthermore, the melting rate influences the liquid pool shape and depth, which have a direct effect on the ingot solidification structure. From a modeling point of view, it is obvious that the numerical simulations must include the melting stage of the process so that numerical predictions can be strictly correlated to operating conditions such as electrical parameters without the need to specify the melt rate value as an input data of the simulation.

Thanks to the use of a well-dedicated electron beam laboratory furnace, a campaign of drip melting trials was performed with the aim of investigating the thermal behavior of a Ti-64 bar. Surface temperature of the bar tip was measured by pyrometer and infrared thermocamera, and the melting rate was recorded. As part of our modeling effort of the EBM processes, a mathematical model of the thermal behavior of the primary material (load) during melting has been developed. Details of this model are given in the present paper. The results of the model are compared to melt rate experimental data obtained with the laboratory EBM furnace, so that clear conclusions can be drawn.

2. Materials and Methods

2.1. Experimental Setup and Procedure

A set of experiments was conducted in a 100 kW laboratory Electron Beam Melting furnace (ALD Vacuum Technologies—Lab100, Hanau, Germany) shown in Figure 2. The EB gun was used at an electric power from 11 kW to 25 kW with a constant beam voltage of 40 kV. The beam path is controlled by the Escosys pilot program (ALD Vacuum Tecnologies, Hanau, Germany), which allows automatic control of beam displacement, as well as a choice of beam pattern shape and frequency. A FLIR X6540SC infrared (IR) camera (Teledyne FLIR Systems, Thousand Oaks, CA, USA) with a wavelength of 2–5 µm was set up at one side of the furnace to evaluate the temperature of the bar tip as well as qualitative IR pictures of the bar side. This camera provides temperature measurements from low ambient temperature to beyond 2000 °C with an uncertainty of ±2% of measured values. The view port was equipped with a BaF$_2$ glass transparent to infrared radiation. Since the glass view was the object of a possible quick coating with vaporized titanium and aluminum, a flow of argon in front of the viewpoint shutter was applied to prevent deposition. A two-wavelength pyrometer (IRCONE Modline Type-R, FLUKE Corp, Everett, WA, USA) was used to measure the surface temperature and to calibrate the infrared camera.

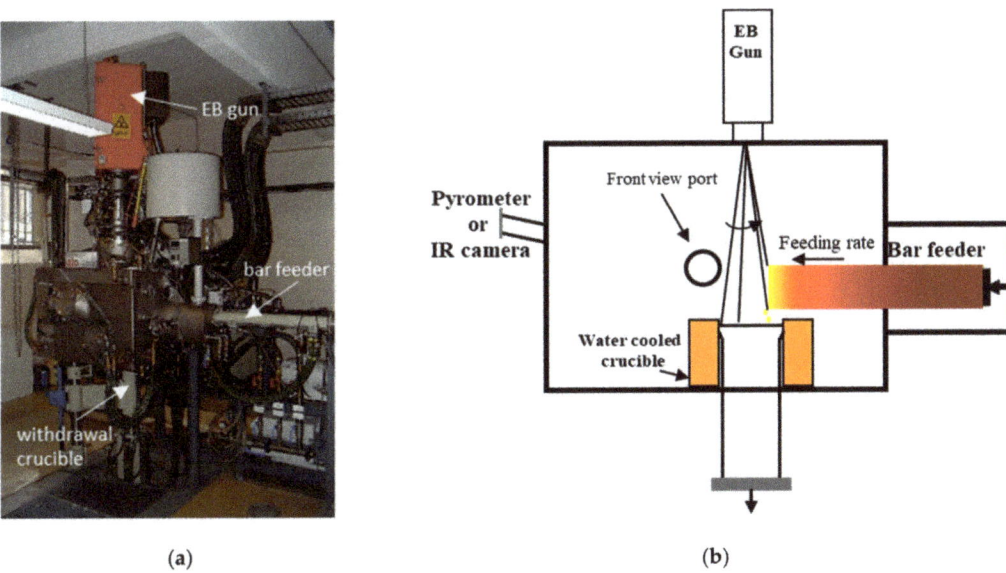

Figure 2. (**a**) Picture of the EB-Lab100 furnace; (**b**) schematic diagram of the Ti-64 bar melting.

The Ti-64 bar (45 cm in length and with a section of 25.0 cm^2) was placed in the bar feeder and the EB chamber vacuumed to below 5×10^{-4} mbar. The EB gun was then turned on, a first pattern being used to heat the bar tip while the ingot surface was heated by a second pattern. A few moments were spent in each trial on finding the exact required power of the beam for a steady melt according to the bar feeder speed. Initially, both the bar and ingot feeder speeds were changed manually; then, the speed of the ingot feeder was adjusted according to melting rate. Later, the bar and ingot feeder rates were set on automatic so that, when the quasi-stationary state is reached, the bar tip is maintained at the same location above the ingot crucible. During this quasi-stationary melting rate period temperature measurements using both pyrometer and IR camera were performed. After the consumption of 90% of the bar length, the beam was switched off and materials were cooled down. Finally, the post-experimental microstructure of the bar tips was observed and compared for different trials.

2.2. Numerical Modeling

This section describes the numerical modeling of the transient thermal behavior of the consumable Ti-64 bar in the EBDM process. An original feature of the model is the representation of liquid metal removal at the bar tip, which allows the model to predict the evolution of the melt rate with time, for given operating parameters.

2.2.1. Heat Transfer Model
Model Equations

The modeling of the thermal behavior of the consumable bar is developed in a 2D axisymmetric geometry, enabling a representation of the curvature of the bar tip. Within the solid metal, heat is transferred by conduction. Under these conditions, the heat transport equation is written in the following form [8]:

$$\rho C_P \frac{\partial T}{\partial t} = \nabla \times \left(\lambda \vec{\nabla} T\right) \quad (1)$$

where T is the temperature, ρ the density, C_P the specific heat and λ the thermal conductivity. The temperature dependence of the alloy thermophysical properties is taken into account.

Dissipation of the latent heat of melting is represented using the equivalent specific heat method [9], which consists of replacing the specific heat in Equation (1) by

$$C_P^* = C_P + \frac{\partial g_l}{\partial T} L \quad (2)$$

where L is the latent heat of melting, and g_l is the liquid fraction.

The calculation of the equivalent specific heat requires the knowledge of the solidification path, i.e., the evolution of the liquid fraction with temperature. In the case of a multicomponent alloy (such as Ti-64), the solidification path is in general unknown. Under these conditions and because the interval between liquidus and solidus is very small, we consider a linear variation of the liquid fraction between the solidus (T_{sol}) and liquidus (T_{liq}) temperatures. Such a variation implies the assumption of a uniform dissipation of the latent heat L in the temperature interval [T_{sol}, T_{liq}], which is particularly narrow in the case of Ti-64. Equation (2) can thus be written as

$$C_P^* = C_P + \frac{L}{T_{liq} - T_{sol}} \quad (3)$$

Initially, the bar temperature is considered to be homogeneous, equal to the stub or pusher temperature (T = T_0) i.e., the room temperature. The symmetry condition on the bar axis is given by $\left(\frac{\partial T}{\partial r}\right)_{r=0} = 0$. The boundary conditions on the edge and the tip of the bar depend on the process itself.

Boundary Conditions

At the bar tip, the kinetic energy of the electron beam is converted into thermal energy; however, the backscattering of electrons results in significant losses, which are in the range of 30% of the incident energy [10] for titanium. Thus, the heat power density applied to the tip surface of the bar is given by:

$$\varphi_{EB} = \frac{\varepsilon_{EB} P_{EB-tip}}{\pi R^2} \quad (4)$$

where P_{EB-tip} is the electrical power of the beam applied to the tip, and ε_{EB} is equal to 0.7. In addition, the net heat exchanged by radiation between the liquid film and the furnace wall must be taken into account. Since the surface ratio between the bar and the furnace wall is very small and in the approximation of the grey body, the thermal flux density transferred can be expressed as:

$$\varphi_{rad}^{tip} = \sigma \varepsilon_l \left(T^4 - T_w^4\right) \quad (5)$$

where T_w is the temperature of the furnace wall.

At the bar lateral surface, heat transfer is controlled by thermal radiation between the material and the furnace walls with a similar expression as Equation (5):

$$\varphi_{rad}^{side} = \sigma \varepsilon_s \left(T^4 - T_w^4\right) \quad (6)$$

At the end of the bar, a contact resistance between the stub or pusher and the bar is taken into account using a heat transfer coefficient h_{sp}. This coefficient is assigned a constant value of 500 W.m^{-2}·K^{-1}, describing a moderately effective thermal contact between the pusher and the bar [11,12].

$$\varphi_{push} = h_{sp}\left(T - T_{push}\right) \qquad (7)$$

The integration of these flux densities (Equations (4)–(7)) over the respective surface leads to heat power defined as $\varepsilon_{EB} P_{EB-tip}$, $\dot{Q}_{rad}^{tip}$, $\dot{Q}_{rad}^{side}$, $\dot{Q}_{push}$ for respectively the effective EB heat power, heat lost by radiation at the tip, heat lost by radiation at the bar side and heat lost at the pusher contact.

The thermal properties of Ti-64 used in the model are reported in Table 1. Note that the thermal emissivity of the liquid titanium has been obtained by IR and pyrometer measurements (see Section 3.2).

Table 1. Thermal properties of Ti-64 used in the numerical simulations [13,14].

Properties of Ti-64	Values or Expression
ρ_l liquid density (kg·m^{-3})	4100
ρ_s solid density (kg·m^{-3})	4158
T_{liq} liquidus (°C)	1670
T_{sol} solidus (°C)	1650
L latent heat of melting (J·kg^{-1})	3.89×10^5
C_p specific heat of solid (J·kg^{-1}·K^{-1})	710
C_p specific heat of liquid (J·kg^{-1}·K^{-1})	794
λ_s conductivity of solid (W·m^{-1}·K^{-1})	18.4
λ_l conductivity of liquid (W·m^{-1}·K^{-1})	22.5
ε_l emissivity of liquid	0.22
ε_s emissivity of solid	0.43 [14]

2.2.2. Calculation Procedure

The heat transfer equation is solved using a finite volume method [15]. The numerical program called 'Ebmelting' is written in FORTRAN. A typical 160 × 1500 (r,z) orthogonal grid is applied for the spatial discretization. Furthermore, a fully implicit scheme is used for time discretization. During each time step (typically 0.5 s), we first calculate the temperature field in the material. Then, in order to simulate consumption of the bar associated to the fall of liquid metal droplets formed at the bar tip, the mesh cells whose temperature is greater than an "overheating temperature" T_{oh} are removed from the computational domain. After mesh cell removal, the boundary conditions are set at the new bar tip for the next time step. Calculations are performed until full consumption of the bar or until a given processing time. Note that the overheating temperature T_{oh} is an input variable for Ebmelting and is defined as the sum of the alloy liquidus temperature and a superheat.

Finally, Ebmelting calculates at each time increment the temperature field of the bar, the shape of the bar tip as well as the melting rate.

3. Results and Discussion

3.1. Experimental Melting Rate and Overheating Temperature

A set of eight trials has been set up in our laboratory electron beam furnace with associated melting rate and EB power as reported in Table 2. Notice that the total EB power P_{EB-tot} has two contributions, one is the electrical power used to scan the pattern at the bar tip P_{EB-tip} and the second is the power P_{EB-ing} applied to the ingot top for solidification control as schematically described in Figure 2b.

$$P_{EB-tot} = P_{EB-tip} + P_{EB-ing} \qquad (8)$$

Table 2. Experimental values for each individual run.

Run #	Bar Feeder Speed (mm·min⁻¹)	Melting Rate $\dot{m}_m$ (kg/h)	$P_{EB\text{-}tip}$ (kVA)	$P_{EB\text{-}tot}$ (kVA)	Measured Overheating Temperature (°C)
1	8	5.15	5.81	11.6	1760
2	10	6.44	6.38	12.4	1770
3	12	7.73	7.65	13.6	1762
4	15	9.66	8.33	14.8	1760
5	20	12.88	9.68	17.2	1772
6	25	15.90	12.16	21.6	1772
7	27	18.58	13.06	23.2	1784
8	30	20.65	14.00	24.8	1779

Table 2 clearly reveals that the temperature of the bar tip is not correlated with the EB power applied, and its value remains in a range between 100 and 125 °C above the liquidus temperature. On the contrary, the melting rate increases almost linearly with the heating power as it is shown in Figure 3.

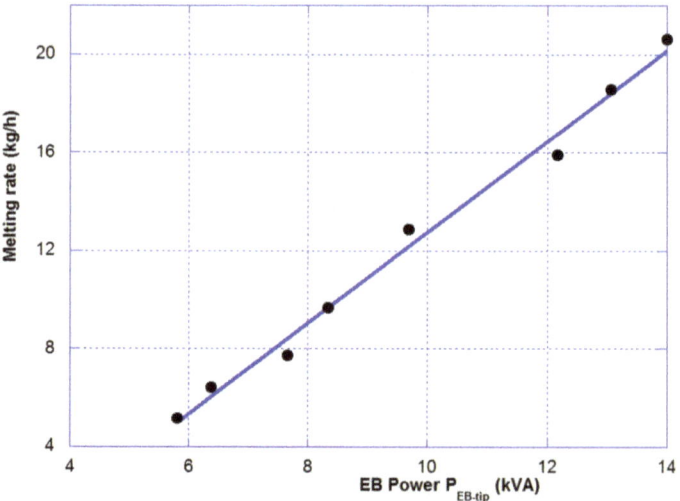

Figure 3. EB power $P_{EB\text{-}tip}$ vs. melting rate. Slope = 1.8 kg/hr/kW.

During the quasi-stationary state, a straightforward heat balance based on the schematic Figure 4 leads to

$$\left(h_{liq}(T_{oh}) - h_{sol}(T_0)\right)\dot{m}_m = \varepsilon_{EB}P_{EB-tip} - \left[\dot{Q}_{rad}^{tip} + \dot{Q}_{rad}^{side} + \dot{Q}_{push}\right] \tag{9}$$

where h_{liq} and h_{sol} are respectively the specific enthalpy of liquid and solid, and $\dot{m}_m$ is the melting rate. The thermal power lost by radiation on the bar tip remains at a constant value (since the overheating temperature T_{oh} does not change with the beam power), and the heat lost on the pusher side can be considered as negligible during the quasi-stationary regime.

Figure 4. Heat fluxes at the bar boundaries.

If the heat lost by radiation on the bar side $\dot{Q}_{rad}^{side}$ was independent of the EB heat power applied on the bar tip, then the melting rate would be proportional to that EB power applied and the slope of the adjusted line would be equal to:

$$S = \varepsilon_{EB}\left(h_{liq}(T_{oh}) - h_{sol}(T_0)\right)^{-1} \qquad (10)$$

The analytical value of S (1.5 kg/hr/kW) does not match with the experimental slope (see Figure 3), which is equal to 1.8 kg/hr/kW. We will see later (Section 3.5) that the temperature profile on the bar side cannot be considered as independent of the beam power. Therefore, the assumption of an independent radiation heat flux on the bar side is wrong.

3.2. Emissivity Measurements and Temperature Profiles

In a first step, the temperature of the bar tip is continuously measured with the IR camera during a cooling test (EB is sharply switched off). The emissivity is then selected such as the measured temperature of the phase change ranges between 1650 °C (T_{sol}) and 1670 °C (T_{liq}). An emissivity of 0.22 was obtained. In a second step, the average surface temperature of the bar tip was monitored by the IR camera during a drip melting run and compared to the temperature measured with the pyrometer. Again, the value 0.22 was required to match IR and pyrometer temperatures. This value is in agreement with the findings of Choi [16] and Rai [17] (0.23 and 0.20, respectively).

An example of IR picture is presented in Figure 5 for a low (Run#1) and a high (Run#5) melting rate. It clearly emphasizes the steeper temperature gradient at high melting rate. The heat supplied by the EB diffuses more deeply into the bar at lower melting rate.

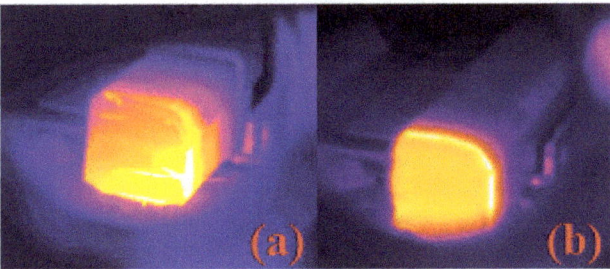

Figure 5. Temperature visualization for two conditions: (**a**) low melting rate—8 mm/min and (**b**) high melting rate—20 mm/min.

3.3. Microstructure of the Bar Tip

In order to get a better understanding of the thermal treatment experienced by the bar during electron beam drip melting (EBDM), the microstructure of the remaining bar tips, after two heats performed at distinguishable melting rates, was studied. In both cases, the melt was sharply interrupted thanks to the EB switch-off. Two samples were obtained from the experiments with a bar feeding speed of 8 and 20 mm/s (melting rate of 5.2 and 12.9 kg/h, respectively). The samples were prepared by cutting in the middle longitudinal section the first tens of millimeters from the front surface of the bar, then polished and etched by Kroll agent. Finally, they were examined under Zeiss-Axioplan 2 optical microscope (Zeiss, Iena, Germany).

Figure 6 illustrates the different grain structure in the first millimeters from the extremity of the bar. The tip of the bar is clearly distinguished on the right-hand side of each image.

(a) (b)

Figure 6. Observed grain structure near the bar tip for (a) low melting rate—8 mm/min and (b) high melting rate—20 mm/min.

The grain morphologies are equiaxed with a size noticeably larger for the lower melting rate. This is caused by the longer time this region was subjected to high temperature (above the beta transus—see below).

From these local images obtained by optical microscopy, mapping pictures of the whole bar tip could be built thanks to the automatic driven system option integrated in Axiovision software (Zeiss, Iena, Germany). Indeed, this system allows to stitch, with a controlled overlap percentage, a large number of images (each square, as the ones dashed in yellow, is a single optical microscope image). These microstructure maps are shown in the Figure 7 for both trials.

On Figure 7, the Heat Affected Zone (HAZ) can be easily unveiled because of microstructure change during heating. According to the Ti-64 phase diagram, beta grain growth occurs during heating at a temperature higher than the beta transus temperature $T_{\alpha\beta}$ (around 880 °C). Because of the higher mobility of beta Ti for grain growth due to its single-phase solution nature whereas the alpha Ti has a lower mobility due to its multiphase nature, this heating above the transus leads to a large grain structure [18] before quenching as electron beam power is turned off.

Consequently, HAZ boundaries have been sketched in dotted line on the images. On the left-hand side of this limit, primary Ti-64 that has not experienced the beta transus can be seen. On the other side, beta grain size enlargement can be noticed from the HAZ beginning to the bar tip. An average value for HAZ was approximately measured for two samples and determined to be around 21 and 11 mm at low and high melting rates, respectively.

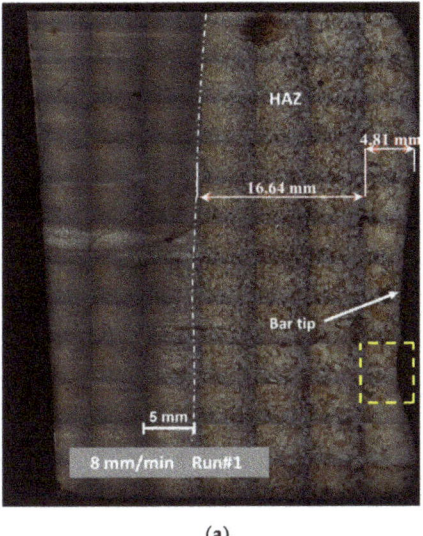 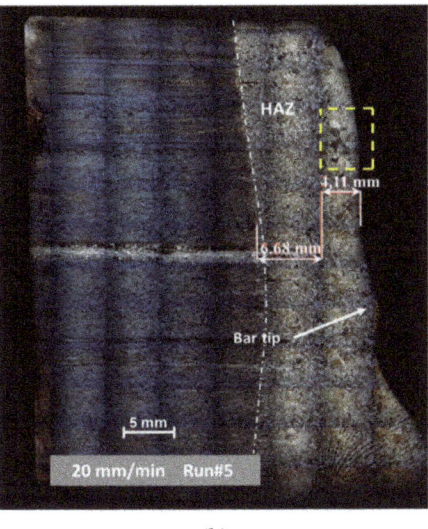

Figure 7. Assessment of the HAZ for (**a**) low melting rate—8 mm/min and (**b**) high melting rate—20 mm/min. Stitched optical microscope images for mapping (unit in dashed yellow).

3.4. Aluminum Depletion by Volatilization

Because of high vacuum, the main drawback of the EBM processes is a significant loss by volatilization of alloying elements exhibiting a high vapor pressure (such as aluminum in Ti-64). This provokes a pollution of the chamber walls and makes it difficult to control the chemical composition of the as-cast ingot.

The significant temperature gradient in the liquid film makes the chemical analysis of the bar tip difficult and inaccurate. This is the reason why the volatilization losses have been obtained with a chemical analysis of the cast ingots using the Glow Discharge Optical Emission Spectrometry (GDOES) technique. Table 3 reports the aluminum content of the initial bar and the mean value of the radial ingot section for the low and high melting rate runs.

Table 3. Aluminum content in the raw material and in the final ingot.

	Al wt%—Initial Bar	Al wt%—Final Ingot
Run#1—8 mm/min	6.0	4.67
Run#5—20 mm/min	6.0	5.39

The analyses emphasize that the EBDM at higher melting rate reduces the losses by volatilization. Since the overheating temperature does not change with the EB power, the volatilization flux on the bar tip remains roughly constant. It is therefore obvious that a higher melting rate reduces the residence time of the alloy in a liquid state and then the aluminum losses. This result agrees well with the literature [19,20] where the detrimental effect of low melting rate on aluminum depletion was established.

3.5. Results of the Numerical Simulation and Discussion
3.5.1. Melting Rate and Overheating Temperature

The eight experimental runs have been simulated, and the excellent convergence of the heat transfer equation lead to a heat balance lower than 0.1%. One of the most interesting results provided by Ebmelting code is the melting rate profile as shown in Figure 8 for the Run#5 at 20 mm/min. Following the EB heating of the bar tip, the first droplets of liquid metal are formed after 30 s. The melting rate increases very steeply and then levels off at a roughly constant value corresponding to the quasi-stationary state. Finally, the melting rate rises again at the end of the consumption of the bar. The oscillation of the melting rate can be readily explained by the ablation of mesh cells, whose temperatures are greater than T_{oh}. The amplitude of the oscillations is then strictly correlated to the mesh refinement.

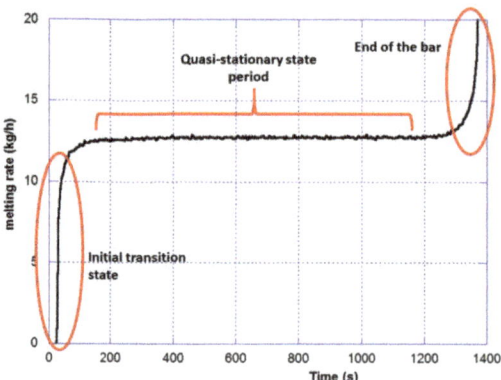

Figure 8. Typical shape of the melting rate (example of Run#5; 20 mm/s)—Mesh refinement (r,z): 160 × 1500.

For each experiment, a wide range of overheating temperature was set into the numerical simulation model as an input to calculate the corresponding melting rates as illustrated in Figure 8. The melting rate calculated by the model obviously decreases with the overheating temperature since a higher thermal energy is required to remove the liquid Ti-64 at the bar tip. Moreover, the heat lost by radiation $\dot{Q}_{rad}^{tip}$ increases non-linearly with T_{oh}, which results in a lower melting rate.

Then, the measured (continuous line) and simulated (markers) melting rates were compared as shown as an example in Figure 9. The intersection of calculated and experimental melting rate gives an assessment of the overheating temperature required in the model.

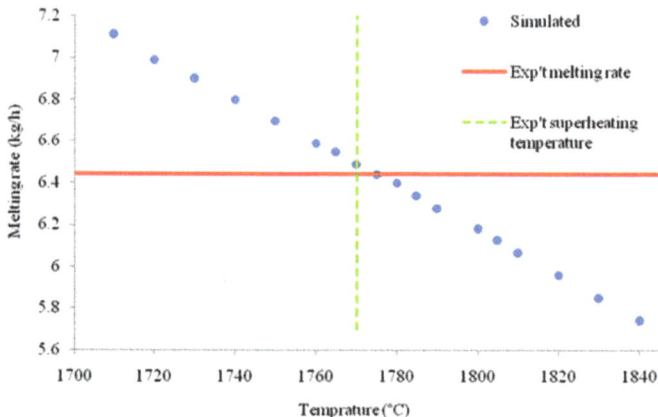

Figure 9. Simulated melting rate versus overheating temperature T_{oh} and measured temperature and melting rate (Run #2: 10 mm/min).

The latter matches well with the measured tip temperature, as shown in Table 4, that gathers the simulated and experimental values for each individual run. A good agreement between measured and indirectly calculated overheating temperatures is obtained in all cases except in Run#3 where the substantial difference could be attributed to an accumulation of experimental errors.

Table 4. Simulated and experimental values for each individual run.

Run #	Bar Feeder Speed (mm/min)	Melting Rate $\dot{m}_m$ (kg/hr)	P_{EB-tip} (kVA)	Experimental Overheating Temperature (°C)	Simulated Liquid Temperature T_{oh} (°C)
1	8	5.15	5.81	1760	1790
2	10	6.44	6.38	1770	1775
3	12	7.73	7.65	1762	1822
4	15	9.66	8.33	1760	1798
5	20	12.88	9.68	1772	1750
6	25	15.90	12.16	1772	1770
7	27	18.58	13.06	1784	1762
8	30	20.65	14.00	1779	1781

Table 4 also confirms that the overheating temperature is not correlated with the melting rate, and its value remains in a range between 90 and 150 °C above the liquidus temperature (T_{liq} = 1670 °C).

3.5.2. Thermal Profiles and HAZ

The numerical model provides at each time step of the simulation the temperature distribution in the bar. Figure 10 presents such computed thermal map under two different melting rates at t = 1000 s. The model predicts that the highest axial temperature gradient is confined close to the bar tip whereas the main part of the bar remains at room temperature. This thermal behavior is easily attributable to the competition between the consumption speed and the thermal diffusion velocity. Calculation of a dimensionless Péclet number allows assessment of the relative role of each of these two phenomena (ratio of advective to diffusive). It gives:

$$Pe_{Run1} = \frac{u_b L}{\alpha} = 7.3 \text{ and } Pe_{Run5} = 27 \qquad (11)$$

where L is the bar length, u_b the bar speed and α the thermal diffusivity. Accordingly, these values testify that on the one hand the bar moves forward much faster than the heat diffuses and, on the other hand, the higher the melting rate, the steeper the temperature gradient.

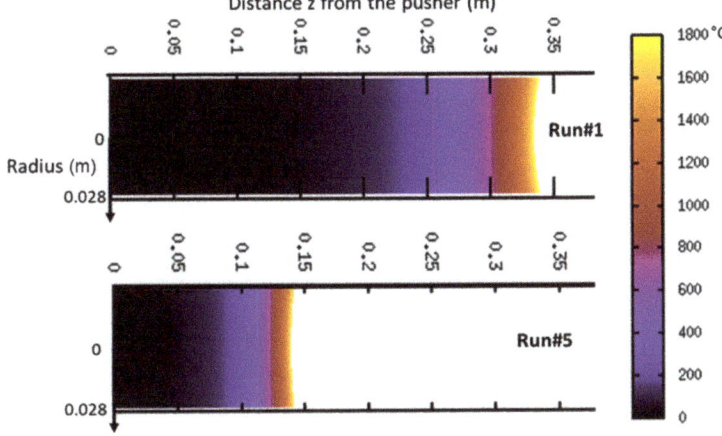

Figure 10. Contour of the temperature after 1000 s of melting Run#1 at 8 mm/min and Run#5 at 20 mm/min.

The concave shape of the bar tip results directly from the thermal losses by radiation along the edge of the bar. This shape remains essentially constant throughout the melt, but we can notice that the curvature of the tip is more pronounced at the beginning of the melt, when EB power is low. In that case, the effect of radiation losses is proportionally more important and Ti-64 melting is, from a thermal point of view, enhanced at the center compared to the periphery. However, the discrepancy with the real shape observed in Figure 7 can be readily explained by the fact that the model does not take into account the dynamic motion of the liquid film under forces such as wetting, buoyancy and thermocapillary [21].

As discussed above, the Heat-Affected Zone (HAZ) is the part of the bar that experiences a temperature over the beta transus $T_{\alpha\beta}$. Profiles of normalized temperature of the bar side are drawn in Figure 11 for the two melting rates (Run#1 and Run#5) and allow determination of the predicted HAZ. The normalized temperature is given by:

$$T^*(z) = \frac{T(z) - T_0}{T(z=0) - T_0} \tag{12}$$

where z is here the distance from the bar tip.

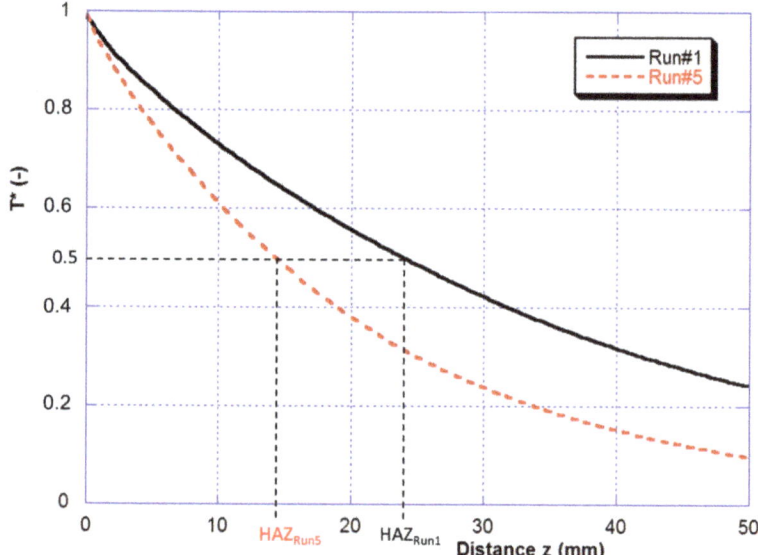

Figure 11. Comparison of the normalized wall temperature profiles in two cases of low and high melting rates (Run#1—8 mm/min and Run#5—20 mm/min).

The computed HAZ values for Run#1 and Run#5 (respectively 24 and 14 mm) match quite well with the measured values (respectively 21 and 11 mm) deduced from the micrographs on Figure 7.

3.5.3. Calculation of Thermal Balance

The global thermal balance applied to the bar was calculated during the quasi-stationary stage by the numerical model Ebmelting, and an example of the results is reported in Table 5. Definition of the variables are given in Section 2.2 except for the heat power requested to heat and melt the material:

$$\dot{Q}_{melting} = \dot{m}_m \left(h_{liq}(T_{oh}) - h_{sol}(T_0) \right) \tag{13}$$

Table 5. Global thermal balance (W) of the bar calculated during quasi-stationary regime.

Watt:	$\varepsilon_{EB} P_{EB-tip}$	$\dot{Q}_{rad}^{tip}$	$\dot{Q}_{rad}^{side}$	$\dot{Q}_{push}$	$\dot{Q}_{melting}$	Balance (%)
Run#1	4068	561	964	0	2542	0.02
Run#5	6780	529	450	0	5808	0.10

The excellent convergence of the simulation is proved by the small balance (lower than 0.1%). The heat lost at the pusher side remains negligible as long as the bar length is much larger than the HAZ. Only during the last 100 s of the melting, $\dot{Q}_{push}$ is no longer negligible. Finally, as a consequence of the bar consumption, it is found that the heat lost by radiation clearly decreases when the melting rate increases, as discussed in Section 3.1.

4. Conclusions

Experiments of melting Ti-64 bars in a laboratory electron beam furnace were compared to a numerical model accounting for the various heat transfers in the titanium bar and at its surfaces. The experiments were designed to mimic the melting step in EBDM and EBCHM processes in order to investigate the effect of operating conditions in such processes. Melting rate and beam power heating the bar tip are adjusted altogether to achieve a quasi-steady melt. After experimentally determining the emissivity of liquid titanium, the overheating temperature at the bar tip was measured for 8 sets of operating parameters. These temperature measurements supported by the numerical results provided the undisputed evidence that the overheating temperature remains roughly independent of the melting rate (and therefore the EB power applied to the bar tip) and equal to about 100 °C above the liquidus temperature. This important finding indicates that high melting rate operations favor the control of the chemical composition of the re-melted material, notably in terms of aluminum depletion by volatilization. The heat affected zone in the bar was optically determined from the grain structure after quenching for two of these sets that differ significantly (melting rate more than doubled). While the overheating temperature is constant, the HAZ was found to strongly depend on operating parameters. The predictions of the numerical model for overheating temperature and HAZ are in good agreement with the experimental measurements, which validates the hypotheses on which the model is built.

Numerical simulations and experiments together demonstrate that the consumption speed of the raw material is much faster than the thermal diffusion velocity, and as a consequence, the higher the melting rate, the steeper the temperature gradient and the smaller the HAZ.

Overall, the experimental results on the melting of a titanium bar by EBM provide knowledge on this step that conditions industrial processes such as EBDM and EBCHM, but that was only scarcely studied in the scientific literature. The associated numerical model not only provides a way to quantify the respective contributions of several heat fluxes in the process, but since it has been validated against experiments, it also provides a relatively simple yet predictive model of EBM that can be used as input in further studies of the downstream steps that distinguish EBDM and EBCHM processes.

Author Contributions: Conceptualization, J.-P.B. and A.J.; methodology, J.-P.B.; software, A.J.; validation, J.-P.B., J.J. and A.J.; formal analysis, J.-P.B., J.-S.K.-R., T.Q. and A.J.; experimental investigation, J.J.; writing—original draft preparation, J.-P.B.; writing—review and editing, J.-P.B., A.J., J.-S.K.-R., and T.Q. All authors have read and agreed to the published version of the manuscript.

Funding: This research received no external funding.

Institutional Review Board Statement: Not applicable.

Informed Consent Statement: Not applicable.

Data Availability Statement: The data presented in this study are available on request from the corresponding author.

Conflicts of Interest: The authors declare no conflict of interest. The funders had no role in the design of the study; in the collection, analyses or interpretation of data; in the writing of the manuscript or in the decision to publish the results.

References

1. Sandell, V.; Hansson, T.; Roychowdhury, S.; Mansson, T.; Delin, M.; Akerfeldt, P.; Antti, M.L. Defects in Electron Beam Melted Ti-6Al-4V: Fatigue Life Prediction Using Experimental Data and Extreme Value Statistics. *Materials* **2021**, *14*, 640. [CrossRef] [PubMed]
2. Bellot, J.P.; Foster, B.; Hans, S.; Hess, E.; Ablitzer, D.; Mitchell, A. Dissolution of hard-alpha inclusions in liquid titanium alloys. *Met. Trans. B* **1997**, *28*, 1001–1010. [CrossRef]
3. Cen, M.J.; Liu, Y.; Chen, X.; Zhang, H.W.; Li, Y.X. Calculation of flow, heat transfer and evaporation during the Electron Beam Cold Hearth Melting of Ti-6al-4V Alloy. *Rare Met. Mat. Eng.* **2020**, *49*, 833–841.
4. Bellot, J.P.; Dussoubs, B.; Reiter, G.; Flinspach, J. A comprehensive numerical modelling of electron beam cold hearth refining and ingot consolidation of Ti alloys. *Rare Met. Mat. Eng.* **2006**, *35* (Suppl. 1), 93–96.
5. Vutova, K.; Vassileva, V.; Koleva, E.; Georgieva, E.; Mladenov, G.; Mollov, D.; Kardjiev, M. Investigation of electron beam melting and refining of titanium and tantalum scrap. *J. Mater. Process Technol.* **2010**, *210*, 1089–1094. [CrossRef]
6. Bertram, L.A.; Zanner, F.J. Electrode tip melting simulation during vacuum arc remelting of Inconel 718. In Proceedings of the Modeling and Control of Casting and Welding Processes, Santa Barbara, CA, USA, 12 January 1986; p. 95.
7. Jardy, A.; Falk, L.; Ablitzer, D. Energy exchanges during vacuum arc remelting. *Ironmak. Steelmak.* **1992**, *19*, 226–232.
8. Carslaw, H.S.; Jaeger, J.C. *Conduction of Heat in Solids*; Oxford University Press: Oxford, UK, 1959.
9. Zanner, F.J.; Bertram, L.A. Vacuum arc remelting: An overview. In Proceedings of the 8th International Conference on Vacuum Metallurgy, Linz, Austria, 30 October–4 November 1985; Volume 1, pp. 512–552.
10. Shiller, S.; Heisig, U.; Panzer, S. *Electron Beam Technology*; Publ. John Wiley&Sons: New York, NY, USA, 1982.
11. Clites, P.G.; Beall, R.A. *A Study of Heat Transfer to Water-Cooled Copper Crucibles during Vacuum Arc Melting*; Report No. 7035; United States Bureau of Mines: Albany, OR, USA, 1967.
12. Khasin, G.A.; Bigashev, V.Z.; Ermanovich, N.A. Contact phenomena at bottom of an ingot in ESR and VAR processes. *Steel USSR* **1973**, *3*, 383–385.
13. Boivineau, M.; Cagran, C.; Doytier, D.; Eyraud, V.; Nadal, M.; Wilthan, B.; Pottlacher, G. Thermo-physical properties of solid and liquid TA6V titanium alloy. *Int. J. Thermophys.* **2006**, *27*, 507–529. [CrossRef]
14. Shur, B.A.; Peletskii, V.E. The Effect of Alloying Addition on the emissivity of Titanium in the Neighborhood of Polymorphous Transformation. *J. High Temp.* **2004**, *42*, 414–420. [CrossRef]
15. Patankar, S.V. *Numerical Heat Transfer and Fluid Flow*; Hemisphere Publishing Corporation: Washington, DC, USA, 1980.
16. Choi, W.; Jourdan, J.; Matveichev, A.; Jardy, A.; Bellot, J.P. Kinetics of Evaporation of Alloying Elements under Vacuum: Application to Ti alloys in Electron Beam Melting. *High Temp. Mater. Process.* **2017**, *36*, 815–823. [CrossRef]
17. Rai, R.; Burgardt, P.; Milewski, J.O.; Lienert, T.J.; DebRoy, T. Heat transfer and fluid during electron beam welding of 21Cr-6Ni-9Mn steel and Ti-6Al-4V alloy. *J. Phys. D Appl. Phys.* **2009**, *42*, 1–12. [CrossRef]
18. Kobryn, P.A.; Semiatin, S.L. Microstructure and texture evolution during solidification processing of Ti-6Al-4V. *J. Mater. Process. Technol.* **2003**, *135*, 330–339. [CrossRef]
19. Bellot, J.P.; Hess, E.; Ablitzer, D. Aluminium volatilization and inclusion removal in the electron beam cold hearth melting of Ti alloys. *Met. Mater. Trans. B* **2000**, *31B*, 845–854. [CrossRef]
20. Westterberg, K.W.; McClelland, M.A. *Modeling of Material and Energy Flow in an EBCHR Casting System*; No. UCRL-JC-118172, CONF-9411237-1); Bakish, R., Ed.; Lawrence Livermore National Lab.: Livermore, CA, USA, 1994; pp. 185–201.
21. Bhar, R.; Jourdan, J.; Descotes, V.; Jardy, A. An experimental study of the inclusion behavior during maraging steel processing. *Met. Res. Technol.* **2019**, *116*, 517.

Article

Nano-Mechanical Properties and Creep Behavior of Ti6Al4V Fabricated by Powder Bed Fusion Electron Beam Additive Manufacturing

Hanlin Peng [1], Weiping Fang [1,*], Chunlin Dong [1], Yaoyong Yi [1], Xing Wei [2], Bingbing Luo [1] and Siming Huang [3]

[1] Guangdong Provincial Key Laboratory of Advanced Welding Technologies, China-Ukraine Institute of Welding, Guangdong Academy of Sciences, Guangzhou 510650, China; henryhlpeng@163.com (H.P.); dongchl@gwi.dg.cn (C.D.); yiyy@gwi.dg.cn (Y.Y.); lbingbing11@126.com (B.L.)
[2] School of Materials Science and Engineering, Huazhong University of Science and Technology, Wuhan 430074, China; weixing_hust@163.com
[3] School of Mechanical and Automotive Engineering, South China University of Technology, Guangzhou 510640, China; scutsiminghuang@163.com
* Correspondence: fwpln@163.com; Tel.: +86-020-61086345

Citation: Peng, H.; Fang, W.; Dong, C.; Yi, Y.; Wei, X.; Luo, B.; Huang, S. Nano-Mechanical Properties and Creep Behavior of Ti6Al4V Fabricated by Powder Bed Fusion Electron Beam Additive Manufacturing. *Materials* **2021**, *14*, 3004. https://doi.org/10.3390/ma14113004

Academic Editor: Katia Vutova

Received: 28 April 2021
Accepted: 26 May 2021
Published: 1 June 2021

Publisher's Note: MDPI stays neutral with regard to jurisdictional claims in published maps and institutional affiliations.

Copyright: © 2021 by the authors. Licensee MDPI, Basel, Switzerland. This article is an open access article distributed under the terms and conditions of the Creative Commons Attribution (CC BY) license (https://creativecommons.org/licenses/by/4.0/).

Abstract: Effects of scanning strategy during powder bed fusion electron beam additive manufacturing (PBF-EB AM) on microstructure, nano-mechanical properties, and creep behavior of Ti6Al4V alloys were compared. Results show that PBF-EB AM Ti6Al4V alloy with linear scanning without rotation strategy was composed of 96.9% α-Ti and 2.7% β-Ti, and has a nanoindentation range of 4.11–6.31 GPa with the strain rate ranging from 0.001 to 1 s^{-1}, and possesses a strain-rate sensitivity exponent of 0.053 ± 0.014. While PBF-EB AM Ti6Al4V alloy with linear and 90° rotate scanning strategy was composed of 98.1% α-Ti and 1.9% β-Ti and has a nanoindentation range of 3.98–5.52 GPa with the strain rate ranging from 0.001 to 1 s^{-1}, and possesses a strain-rate sensitivity exponent of 0.047 ± 0.009. The nanohardness increased with increasing strain rate, and creep displacement increased with the increasing maximum holding loads. The creep behavior was mainly dominated by dislocation motion during deformation induced by the indenter. The PBF-EB AM Ti6Al4V alloy with only the linear scanning strategy has a higher nanohardness and better creep resistance properties than the alloy with linear scanning and 90° rotation strategy. These results could contribute to understanding the creep behavior of Ti6Al4V alloy and are significant for PBF-EB AM of Ti6Al4V and other alloys.

Keywords: titanium alloys; electron beam additive manufacturing; nanoindentation; strain rate sensitivity; creep

1. Introduction

In recent years, additive manufacturing (AM) has attracted extensive attention and has great potential to be used in the fields of aerospace, automotive, energy, and medicine [1–3]. Powder bed fusion with electron beam (PBF-EB) AM is one of the AM methods, using a high-energy electron beam as a heat source to melt metal powders and then obtaining full-dense metallic components [4]. Compared with the selective laser melting (SLM) AM method, PBF-EB AM has several advantages [5]: (1) High vacuum condition is favorable for materials containing active elements; (2) high energy input is favorable for refractory materials. It has been reported that PBF-EB AM has been applied in fabricating high-entropy alloys [6], titanium alloys [7], and ceramics [8].

Due to their superior mechanical properties, excellent corrosion resistance, and outstanding biocompatibility, Ti6Al4V alloys have been used as biomedical implants and aviation materials [9–12]. However, the inherent difficulties in plastic processing and specific components with complex geometry make Ti6Al4V an attractive material for AM. The abrasion, fatigue, or impact properties of Ti6Al4V alloys could be further improved through AM [13,14]. In comparison with other AM methods, PBF-EB AM is the most widely-used process for producing Ti-based components because of their extremely high sensitivity to oxidation at high fabrication temperatures [15–18]. For AM Ti6Al4V alloys, the most attention is focused on the effects of PBF-EB AM parameters, post-treatment on microstructure, and mechanical properties. Tan et al. [19] investigated the columnar grain growth behavior of Ti6Al4V alloy during the PBF-EB AM process. Leon et al. [9] improved the mechanical properties of PBF-EB AM Ti6Al4V alloys using hot isostatic pressing post-treatment. Zheng et al. [20] investigated the effects of powder usage numbers on the hardness of PBF-EB AM Ti6Al4V alloys.

To date, only a few studies have been done to comprehensively assess the creep behavior of Ti6Al4V alloys [21]. However, the creep behavior is not completely understood. Investigation of creep behavior by traditional uniaxial tensile tests is time-consuming for lots of materials [22]. Nanoindentation tests are confirmed to be effective in analyzing the time-dependent plastic deformation of aluminum [23,24], high-entropy alloys [25,26], titanium alloys [27], and Ni_3Al [28]. Although there is a bit of discrepancy between the indentation creep results and conventional uniaxial results [29–31], creep information including strain rate sensitivity, the creep stress exponent, and creep rate could be obtained based on the nanoindentation creep test [30,32]. Shen et al. claimed the discrepancy between nanoindentation and uniaxial methods results from the deformation mechanics [33]. The response from indentation creep tests thus includes the transient stage as well as the steady-state, or even post steady-state stages, which is more complex than that from uniaxial tests.

Wu et al. [34] found that the scanning strategy during selective laser melting AM has a great influence on residual stress of Ti6Al4V alloys. Tian et al. [35] reported that microstructure and mechanical properties of Ti6Al4V alloys could be influenced by the scanning strategy during selective laser melting AM. However, the reports on the effects of the scanning strategy during PBF-EB AM on nano-mechanical properties and creep behavior of Ti6Al4V alloys are rarely seen. Therefore, in this study, the effects of the scanning strategy during PBF-EB AM on microstructure, nanohardness, strain rate sensitivity, and creep behavior of Ti6Al4V are investigated.

2. Materials and Methods

2.1. Powder Preparation

Spherical Ti6Al4V powders with an average size of 76 μm were provided by the supplier, Guangzhou Sailong Additive Manufacturing Co. LTD., Guangzhou, China, which was fabricated by plasma rotating electrode methods. Powder size distribution was characterized by a laser particle analyzer (MASTERSIZER 3000, Malvern, UK). A scanning electron microscope (SEM, ZEISS, Gemini SEM300, Oberkochen, Germany) equipped with energy-dispersive X-ray spectroscopy (EDS, Bruker, Billerica, MA, USA) was used to evaluate the chemical composition accuracy. The SEM morphology and particle size distribution of the as-received powders are shown in Figure 1a,b, respectively. The corresponding chemical composition was determined to be 5.85 ± 0.4 Al, 4.12 ± 0.24 V (wt.%), and balanced with Ti.

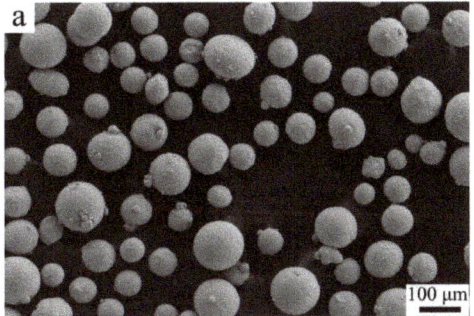 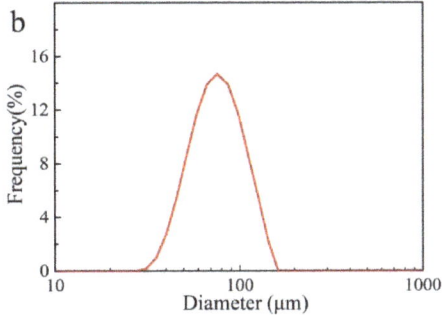

Figure 1. (**a**) SEM morphology and (**b**) particle size distribution of the as-received powders.

2.2. EB-PBF Processing

The cuboid-shaped sample with dimensions of 25 mm × 100 mm × 50 mm was produced using a T150 powder bed fusion electron beam additive manufacturing machine (Guangzhou Sailong Additive Manufacturing Co. LTD., Guangzhou, China). Plasma rotating electrode Ti6Al4V powders (mean diameter of 76 µm) were applied at a layer thickness of 50 µm. During the PBF-EB additive manufacturing, a scan speed of 5800 mm/s, a high voltage of 60 kV, a current of 14.5 mA, and a space of 100 µm between scan lines were adopted. The preheating temperature was kept at 730 °C for all of the samples to avoid powder smoke. Two scanning strategies were used in this study (details can be seen in Figure 2): (1) case (a) is horizontal back and forth linear scanning without rotation on the next layer, and the corresponding sample was referenced as sample A; (2) case (b) is horizontal back and forth linear scanning with a 90° scan vector rotation on the next layer, and the corresponding sample was referenced as sample B.

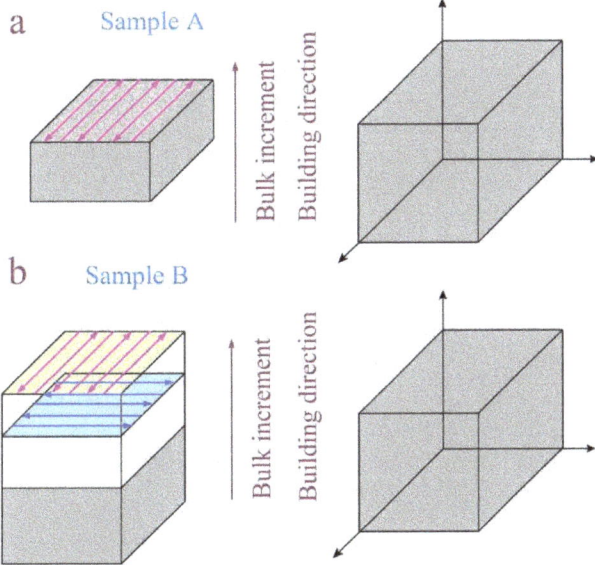

Figure 2. Schematic diagrams of building direction for PBF-EB AM Ti6Al4V alloys: (**a**) Sample A is horizontal back and forth linear scanning without rotation on the next layer; (**b**) Sample B is horizontal back and forth linear scanning with a 90° scan vector rotation on the next layer.

2.3. XRD Analysis

The block samples were cut along the building direction Z-axis using a wire electric discharge machine, 2 mm away from the base, and subsequently embedded for microstructural investigation and nanoindentation tests on the X-Y plane. The cross-sectioned samples were consecutively ground by #100, #600, #800, #1200, #1500, and #2000 grade silicon carbide papers to remove any surface oxides. Then ground samples were consecutively polished with 5 μm, 3 μm, 2 μm, 1 μm, and 0.5 μm grade diamond abrasive paste. Phase constituents were examined by X-ray diffraction (XRD, D/MAX-2500/PC; Rigaku Corp., Tokyo, Japan) with Mo Kα radiation. The angle range of 15–45° and a step size of 0.01° was adopted during the XRD test. More details can be seen in our previous research [36].

2.4. SEM Analysis

Before SEM observation, the polished block samples were etched with Kroll's reagent (2% HF, 6% HNO3, and 92% H_2O) for 5 s. The microstructure of the PBF-EB Ti6Al4V sample was observed by scanning electron microscope (SEM, FEI, QUANT250, Eindhoven, The Netherlands) equipped with energy-dispersive X-ray spectroscopy (EDS), with a working distance (WD) of 9.8 mm and high voltage of 10.00 kV. Additionally, the morphology and chemical composition of the raw powders were observed by scanning electron microscope (SEM, ZEISS, Gemini SEM300, Oberkochen, Germany) equipped with energy-dispersive X-ray spectroscopy (EDS, Bruker, Billerica, MA, USA). After the samples were electrolytically polished with 5% perchloric acid + 95% alcohol and electron backscattered diffraction (EBSD, FEI, Hillsboro, OR, USA) tests were performed to further investigate the microstructure. During the EBSD test, a step size of 0.1 μm was adopted for low magnification, and a step size of 0.01 μm was adopted for high magnification. During the EBSD test, a high voltage of 10.00 kV was used.

2.5. Nanoindentation Analysis

The nanomechanical properties and creep behavior of the PBF-EB AM Ti6Al4V alloys were characterized by nanoindentation tests (Hysitron, TI980, Minneapolis, mN, USA). All the tests were conducted at room temperature and performed with a three-sided Berkovich (CSM Instruments, Peuseux, Switzerland) diamond indenter. For investigation of the strain rate sensitivity, all nanoindentation tests were carried out at the same maximum load (10 mN) and with loading rates of 10, 1, 0.1, and 0.01 mN/s. The indenter was then held at the maximum load for 30 s, which was followed by unloading at a rate of 50 mN/s for all tests. For investigation of creep behavior, the specimen was loaded to different maximum loads, Pmax (10 mN, 20 mN, 50 mN, and 100 mN), at a successive loading rate of 0.5 mN/s and held at Pmax for 500 s.

3. Results

3.1. Microstructure

XRD patterns of PBF-EB AM Ti6Al4V alloys (X-Y plane) arre shown in Figure 3, both for sample A with linear scanning strategy, and sample B with linear and 90° rotation scanning strategy. The XRD pattern confirms the dominant presence of an α-Ti phase, and this result is similar to the previous work [9]. Comparing the XRD results in Figure 3a,b, a tiny difference could be observed. For the XRD pattern of sample A (Figure 3a), the intensity of $(10\bar{1}2)$, $(10\bar{1}3)$, and $(11\bar{2}2)$ diffraction peaks belonging to the α-Ti phase are stronger than those of sample B. This result indicates that the scanning strategy of PBF-EB AM probably leads to a difference in the texture of Ti6Al4V alloys.

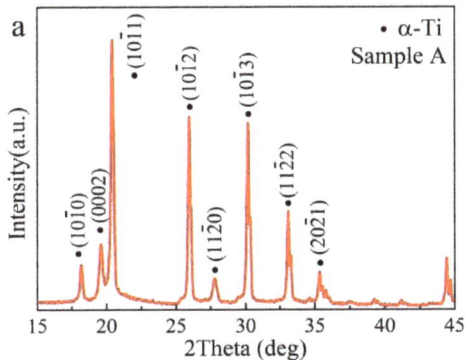 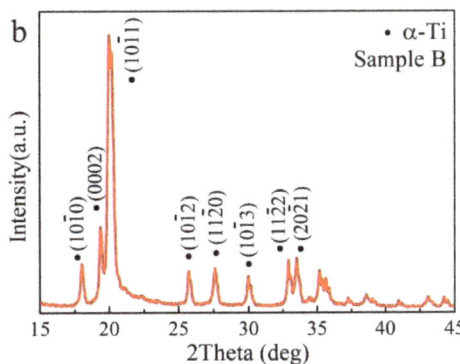

Figure 3. XRD pattern of PBF-EB AM Ti6Al4V alloy: (**a**) linear scanning strategy (sample A), (**b**) linear and 90° rotation scanning strategy (sample B).

As widely known, the microstructure of Ti6Al4V alloy consists of α-Ti and β-Ti phases [37]. During the PBF-EB AM process, the Ti6Al4V alloy originally solidifies into columnar β-Ti grains that grow along the build direction. As the additive manufactured alloys cool to a temperature below around 882 °C, the β-Ti grains boundaries act as nucleation sites for α-Ti grains, and the initial β-Ti within the grains transforms into α/β lamellar structures. This process promotes the development of β-Ti ribs surrounded by a continuous α-Ti phase [38]. The microstructures of sample A and sample B are shown in Figure 4a,b, respectively. It can be seen that the thickness of β-Ti ribs in sample B is smaller than those in sample A. Furthermore, β-Ti ribs in both sample A and sample B have a thickness of few than 1 µm.

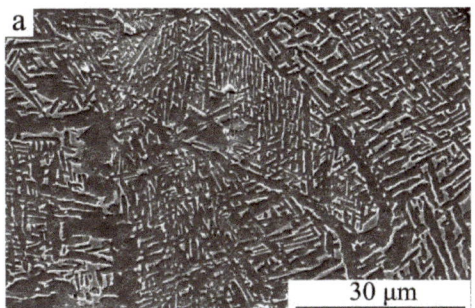 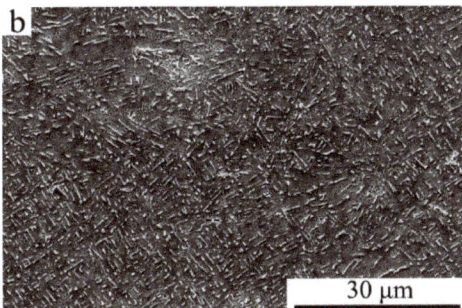

Figure 4. SEM microstructure of PBF-EB AM Ti6Al4V: (**a**) linear scanning strategy (sample A), (**b**) linear and 90° rotate scanning strategy (sample B).

The chemical composition of the a-Ti phase and β-Ti ribs in both sample A and sample B are listed in Table 1. The results are obtained by SEM/EDS. As widely known, Al is an α phase stabilizing element and usually diffused into α phase, while V is a β phase stabilizing element and usually diffused into the β phase [39,40]. It can be seen that there is no obvious enrichment or depth of Al and V elements in the α/β phase. For sample B, both the Al and V concentration in α phase and β ribs is very similar.

Table 1. Chemical composition of α-Ti phase and β-Ti ribs in PBF-EB AM Ti6Al4V with different scanning strategies.

Chemical Composition	Sample A		Sample B	
	α-Ti (wt%)	β-Ti ribs (wt %)	α-Ti (wt %)	β-Ti ribs (wt %)
Al	5.48 ± 0.64	5.86 ± 0.50	5.48 ± 0.07	5.32 ± 0.65
V	4.05 ± 2.55	2.83 ± 0.54	3.22 ± 0.66	2.93 ± 0.66
Ti	90.48 ± 1.92	91.32 ± 0.04	91.2 ± 0.59	90.25 ± 2.14

Since the conditions of heat accumulation and conduction are tightly linked to scanning strategies and relative position of the sample [4,34,41], the X-Y cross-section plane, 2 mm away from the base, was chosen for EBSD observation. Grain morphology and corresponding inverse pole figure (IPF) are shown in Figure 5a,b, respectively. The average grain size of the α-Ti phase is 2.3 µm, as shown in Figure 5c. To analyze the effects of scanning strategy on phase constitution of the PBF-EB AM Ti6Al4V alloy, a region with high magnification and a smaller step size (0.01 µm) was selected. The corresponding results are shown in Figure 5c,d. The phase maps are shown in Figure 5d. It can be seen that the PBF-EB AM Ti6Al4V alloy with only the linear scanning strategy (sample A) was composed of 96.9% α-Ti and 2.7% β-Ti.

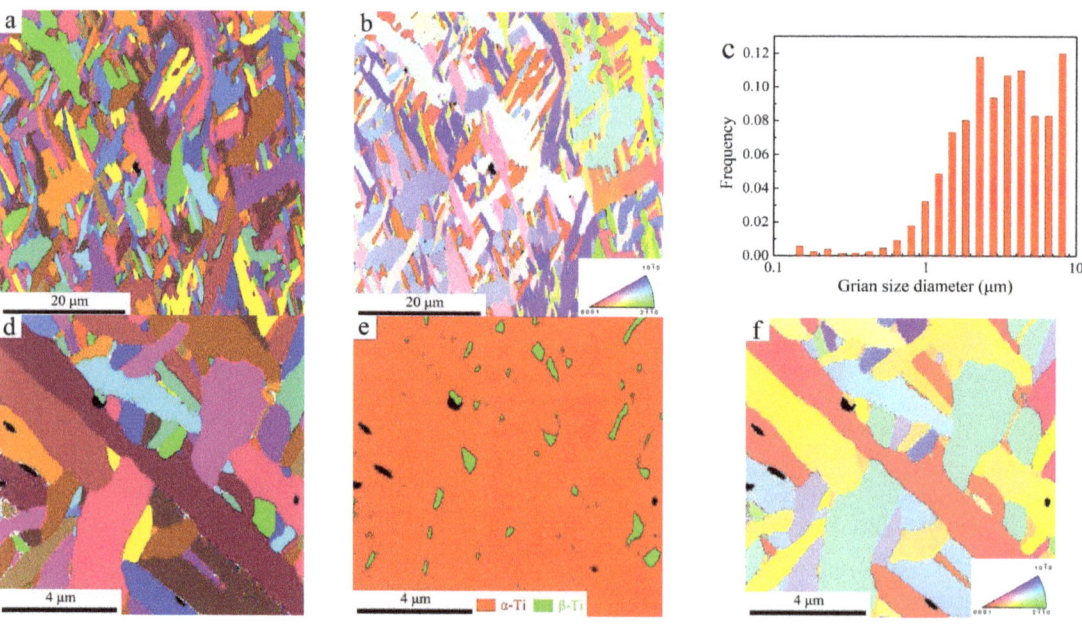

Figure 5. PBF-EB AM Ti6Al4V alloy with only the linear scanning strategy (sample A): (**a**) Grain size and (**b**) IPF images at low magnification; (**c**) grain size, (**d**) phase map, and (**e**) IPF figures at high magnification. The color of the inset figure in (**b**,**f**) represents grains orientation.

For the PBF-EB AM Ti6Al4V alloy with linear and 90° rotation scanning strategy (sample B), the EBSD observation results are shown in Figure 6. From Figure 6c, it can be seen that the average thickness of the α-Ti phase is 2.5 µm. A high magnification was chosen with a step size of 0.01 µm. Additionally, the results are shown in Figure 6c,d. From the phase map (Figure 6e), it can be seen that the PBF-EB AM Ti6Al4V alloy with linear and 90° rotation scanning strategy (sample B) was composed of 98.1% α-Ti and 1.9% β-Ti. Compared with sample B, the α-Ti phase contents in sample A are a bit lower, while the β-Ti phase contents in sample A are a bit higher. The results indicate that scanning strategies during additive manufacturing could affect heat accumulation and conduction, and then influence the microstructure and mechanical properties.

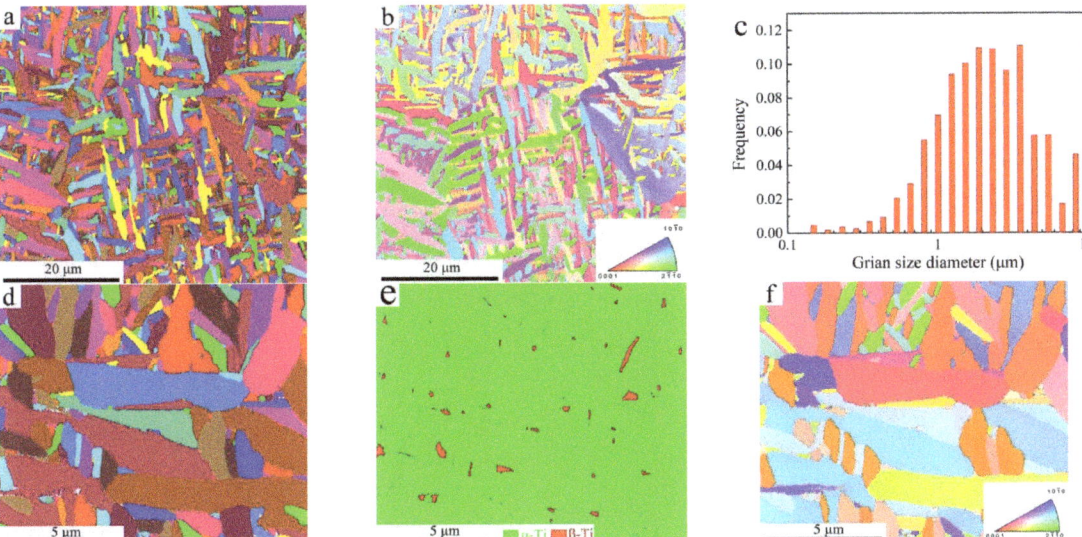

Figure 6. PBF-EB AM Ti6Al4V alloy with linear and 90° rotate scanning strategy (sample B): (**a**) Grain size and (**b**) IPF images at low magnification; (**c**) grain size, (**d**) phase map, and (**e**) IPF figures at high magnification. The color of the inset figure in (**b**,**f**) represents grains orientation.

From the EBSD analysis, the average size of the α-Ti phase in sample A is very similar to the α-Ti phase in sample A. Hence, the grain boundary strengthening is very similar due to the similar thickness of the α-Ti phase [42]. However, the solid solution strengthening is different. From Table 1, it can be seen that Al concentration in the α-Ti phase is similar in both samples. However, the V concentration in the α-Ti phase from sample A is higher than the value from sample B. Therefore, high V concentration in the α-Ti phase would lead to higher hardness.

3.2. Strain-Rate Sensitivity

Before the nanoindentation test, the Archimedes principle was employed to measure the porosity of PBF-EB AM Ti6Al4V alloys. The densities of sample A and sample B were 98.52% and 98.22%, respectively. It can be seen that the scanning strategy has little influence on the porosity of Ti6Al4V alloys under the selected additive manufacturing parameters.

Figure 7 shows the typical load-depth curves for the PBF-EB AM Ti6Al4V alloy as obtained from nanoindentation tests under various loading rates. All nanoindentation tests were carried out at the same maximum load (10 mN). Loading rates of 0.01 mN/s, 0.1 mN/s, 1 mN/s, and 10 mN/s were applied, corresponding to strain rates of 0.001 s^{-1}, 0.01 s^{-1}, 0.1 s^{-1}, and 1 s^{-1}, respectively. As shown in Figure 7, the loading rates result in different depths. The depth increases significantly with the decreasing strain rate (or loading rates). Results from nanoindentation tests under different strain rates are presented in Table 2. It can be seen that as the strain rate increases in the range of 0.001–1 s^{-1}, the hardness increases between 4.11 and 6.31 GPa for sample A. While for sample B, as the strain rate increases in the range of 0.001–1 s^{-1}, the hardness increases between 3.98 and 5.52 GPa. The results indicate that the PBF-EB AM Ti6Al4V alloy with only the linear scanning strategy has higher nanohardness than the alloy with a 90° rotation scanning strategy.

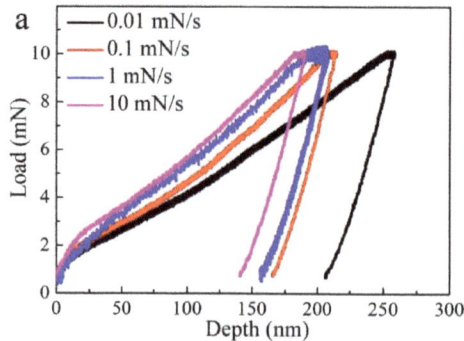 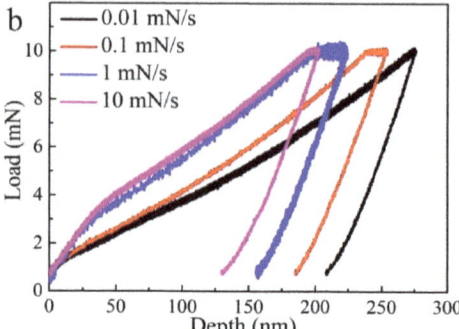

Figure 7. Typical load–depth curve with various strain rates at the same Pmax = 10 mN. (**a**) Linear scanning strategy (sample A), (**b**) linear and 90° rotation scanning strategy (sample A).

Table 2. Various loading rates, strain rates, and corresponding nanohardnesses for PBF-EB AM Ti6Al4V alloys with different scanning strategies.

Loading Rate (mN/s)	Strain Rate (s^{-1})	Hardness (GPa)	
		Sample A	Sample B
0.01	0.001	4.11	3.98
0.1	0.01	5.4	4.33
1	0.1	5.54	4.59
10	1	6.31	5.52

The relationship between indentation nanohardness and strain rate could be expressed by Equation (1):

$$H = C \dot{\varepsilon}^m \quad (1)$$

where H is the nanohardness (GPa), C is the material constant, $\dot{\varepsilon}$ is the strain rate (s^{-1}), and m is the strain rate sensitivity exponent [43]. Therefore, the strain-rate sensitivity exponent can be determined by the slope of an lnH vs. ln $\dot{\varepsilon}$ plot.

The relationship between indentation nanohardness (H) and strain rate ($\dot{\varepsilon}$) is shown in Figure 8. The equations in Figure 8a,b indicate that the relationship between H and $\dot{\varepsilon}$ has a high correlation coefficient, R^2 = 0.83, and R^2 = 0.9 for sample A and sample B, respectively. The results implied that this function can properly measure the dependence of the nanoindentation hardness on the strain rate. Based on the Figure 8, the strain-rate sensitivity exponent (m) are determined to be 0.053 ± 0.014 and 0.047 ± 0.009 for sample A and sample B, respectively.

The strain-rate sensitivity exponent (m) is important to evaluate the super-plasticity of materials. The strain-rate sensitivity exponent (m) of superplastic materials usually exceeds 0.3 [44]. The value of m (0.053 ± 0.014 and 0.047 ± 0.009) in this study does not exceed 0.1, which is similar to that of the most known materials. Yu et al. [43] determined the strain-rate sensitivity exponent of H13 tool steel to be 0.022 using nanoindentation. Jun et al. [27] characterized the strain-rate sensitivity exponent of Ti6246 alloys to be 0.005–0.039 via nanoindentation test.

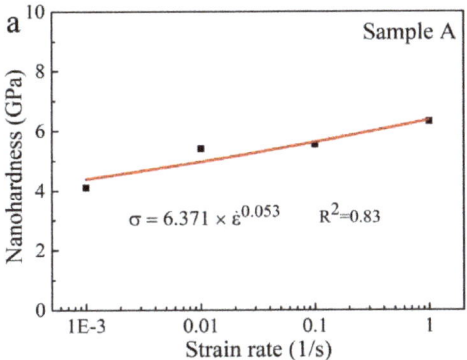 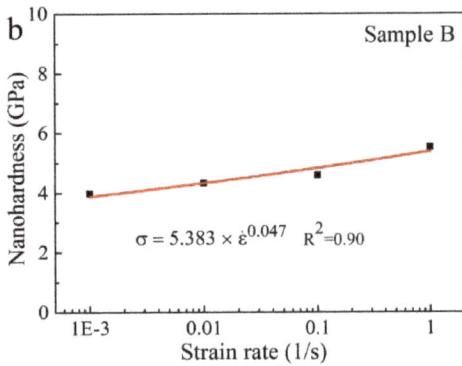

Figure 8. Representative nanohardness as a function of strain rate: (**a**) Linear scanning strategy (sample A) and the slope has a standard error of 0.327, (**b**) linear and 90° rotation scanning strategy (sample B) and the slope has a standard error of 0.181.

3.3. Creep Behavior

For PBF-EB AM Ti6Al4V alloys with different scanning strategies, the typical load–displacement (P–h) curves under various maximum holding loads (Pmax) ranging from 10 to 100 mN are shown in Figure 9a. The curves consist of the loading stage, holding stage (creep stage), and unloading stage. It can be seen that plastic deformation occurred before the holding stage and the displacement increases during the holding stage. With the maximum holding load (Pmax) increased from 10 to 100 mN, the creep displacement dramatically increased. When the unloading is finished, the deformation is not completely recovered and indentation depth remains a large value. The nanoindentation results confirm the occurrence of creep behavior, despite the loading stress being very low at room temperature.

The creep behavior induced by the nanoindentation test has been explored by several scholars, and the creep displacement (h, nm) and holding time (t, s) could be depicted as follows [25,28,32]:

$$h = h_0 + a(t - t_0)^b + kt \quad (2)$$

where h_0, a, t_0, b, and k are constant and can be obtained by fitting the creep displacement (h) and holding time (t, s), h_0 (nm), and t_0 (s) refers to the displacement and time at the beginning stage of creep, respectively. After obtaining the relationship between creep displacement (h, nm) and holding time (t, s), the indentation strain rate ($\dot{\varepsilon}$, s^{-1}) can be obtained according to the equation below [31]:

$$\dot{\varepsilon} = \frac{1}{h}\frac{dh}{dt} \quad (3)$$

where displacement rate (dh/dt, nm/s) could be obtained by derivating the displacement–holding time (h–t). The nanoindentation stress (H, GPa) could be expressed as H = P/h^2, in which P (mN) denotes the holding load and h (nm) refers to the indentation displacement. Then the creep stress exponent (n) could be calculated by the empirical equation below [25]:

$$n = \frac{\partial \ln \dot{\varepsilon}}{\partial \ln H} \quad (4)$$

The experimental variations of displacement with time are shown in Figure 9b, corresponding to the holding stage in Figure 9a. Based on Equation (2), the fitting results of creep displacement (h)–time (t) curves are clearly shown in Figure 9c and a high correlation coefficient (R^2 > 0.9) is obtained. From both experimental and fitting creep displacement-time results, it can be observed that the displacement increased dramatically with increasing loads. The creep displacement has a value of around 15 nm, 30 nm, 75 nm, and 95 nm, corresponding to the maximum holding load, Pmax, of 10 mN, 20 mN, 50 mN, and 100 mN, respectively. According to Equations (2)–(4), the creep stress exponent (*n*) under various indentation loads is shown in Figure 9d. The value of the creep stress exponent (*n*) decreased with the increasing maximum holding loads, Pmax, (10–100 mN).

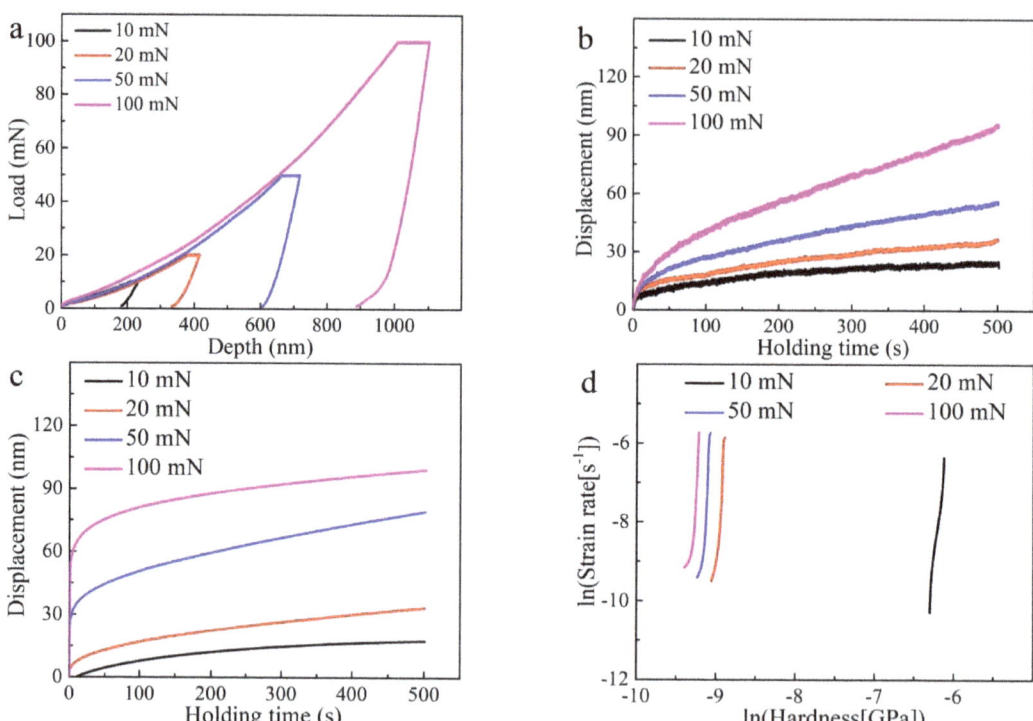

Figure 9. PBF-EB AM Ti6Al4V alloy with only the linear scanning strategy (sample A): (**a**) P–h curves under various maximum loads (10–100 mN); (**b**) experimental and (**c**) fitting creep displacement–time curves under various maximum loads (10–100 mN); (**d**) ln-ln plots of strain rate vs. nanoindentation stress under various maximum loads (10–100 mN).

For PBF-EB AM Ti6Al4V alloy with linear and 90° rotate scanning strategy (sample B), the typical load–displacement (P–h) curves under various peak loads ranging from 10 to 100 mN are shown in Figure 10a. The experimental and fitting creep displacement–time curves under various maximum loads (10–100 mN) are shown in Figure 10b,c, respectively. When fitting the creep displacement–time curves based on Equation (2), there is a high correlation coefficient ($R^2 > 0.9$). When the maximum holding load (Pmax) increased from 10 to 20, 50, and 100 mN, the creep displacement had a values of around 25 nm, 85 nm, 170 nm, and 225 nm, respectively. Figure 10d shows the ln-ln plots of strain rate vs. nanoindentation stress under various maximum loads (10–100 mN). The value of the creep stress exponent (*n*) decreased with the increasing maximum holding loads, Pmax, (10–100 mN).

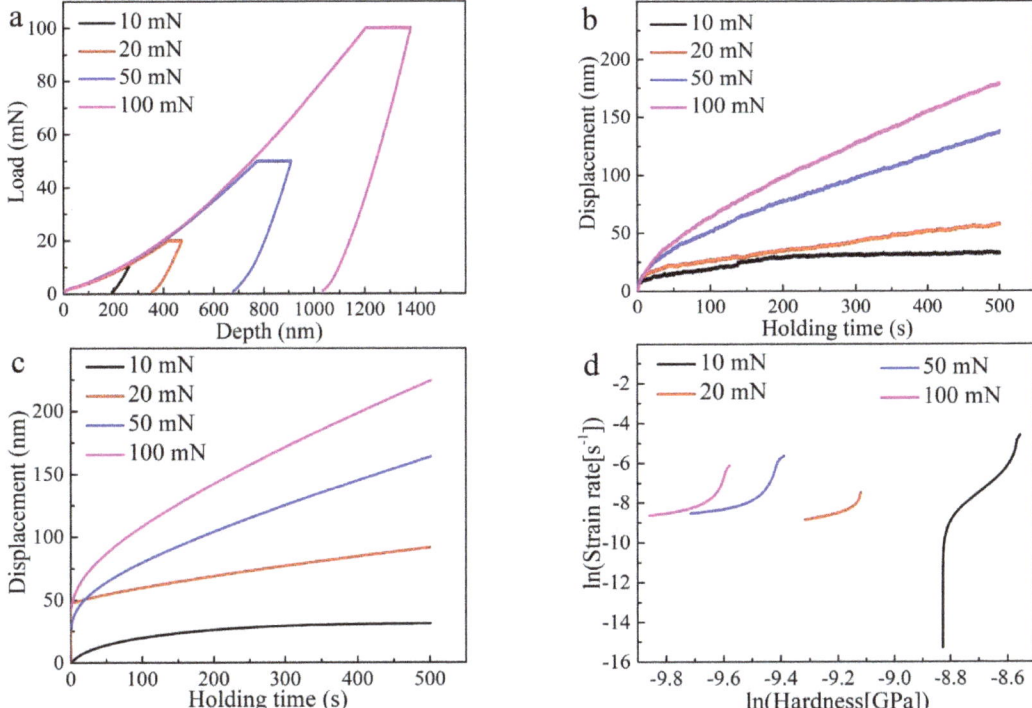

Figure 10. PBF-EB AM Ti6Al4V alloy with linear and 90° rotation scanning strategy (sample B): (**a**) P–h curves under various maximum loads (10–100 mN); (**b**) experimental and (**c**) fitting creep displacement–time curves under various maximum loads (10–100 mN); and (**d**) ln-ln plots of strain rate vs. nanoindentation stress under various maximum loads (10–100 mN).

From both Figure 9b,c and Figure 10b,c, it can be seen that the creep curves from the nanoindentation test have transient-state creep and steady-state creep stages. As widely known, the classic creep curves obtained from uniaxial tensile tests have transient-state creep, steady-state creep, and accelerated creep stages [45,46]. This difference results from the very low loading stress during nanoindentation, which is unable to lead to the catastrophic failure of alloys. At the onset of nanoindentation creep behavior, it has a relatively high strain rate between 10^{-3} and 10^{-4} s^{-1}, which belongs to the transient-sate creep stage. When the holding time is extended, the strain rate decreases to 10^{-4} s^{-1} and enters the steady-state creep stage.

The creep stress exponent (*n*) value is tightly linked to the creep mechanism. Usually, *n* = 1 indicates the diffusion creep mechanism, *n* = 2 indicates the grain boundary sliding mechanism, and *n* = 3–8 indicates the dislocation creep mechanism [23]. For the Ti6Al4V alloy (sample A), the creep stress exponent is calculated to be 39.491 ± 0.324 for 10 mN, 7.213 ± 0.017 for 20 mN, 4.201 ± 0.012 for 50 mN, and 2.556 ± 0.009 for 100 mN, respectively. Additionally, for the PBF-EB AM Ti6Al4V alloy (sample B), the creep stress exponent is calculated to be 11.89 ± 0.02 for 10 mN, 2.666 ± 0.004 for 20 mN, 1.572 ± 0.003 for 50 mN, and 1.842 ± 0.004 for 100 mN, respectively. The creep stress exponent of the PBF-EB AM Ti6Al4V alloy indicates that the creep behavior might be controlled by dislocation motion. The deformation induced by the indenter leads to a high density of dislocation and dislocation motion could effectively occur. Curiously, the calculated creep stress exponents decrease with increasing maximum loads (10–100 mN), which is opposite to the results of CoCrFeMnNi HEAs obtained by He et al. [23]. Lee et al. investigated nanoindentation creep behavior of CoCrFeMnNi high-entropy alloys and obtained creep stress exponent n = 14.34 for Pmax = 10 mN and n = 18.34 for Pmax = 50 mN [41].

4. Conclusions

In this study, the influences of scanning strategy during powder bed fusion electron beam additive manufacturing (PBF-EB AM) on microstructure, nano-mechanical properties, and creep behavior of Ti6Al4V alloy were compared. The results could contribute to understanding the creep behavior of Ti6Al4V alloy and are significant for PBF-EB AM of Ti6Al4V and other alloys. The conclusions can be summarized as follows:

Both PBF-EB AM Ti6Al4V alloys were composed of the predominant α-Ti phase and barely β-Ti phase. Alloys with only the linear scanning strategy were composed of 96.9% α-Ti and 2.7% β-Ti phases, while alloys with linear and 90° rotation scanning strategy were composed of 98.1% α-Ti and 1.9% β-Ti phases. Additionally, the thickness of β ribs in alloys with only the linear scanning strategy are a bit larger than those in the sample with the linear and 90° rotation scanning strategy, but both have a value of lower than 1 μm.

The nanohardness of the PBF-EB AM Ti6Al4V alloy with linear scanning strategy is a bit higher than the value for alloys with linear and 90° rotation scanning strategy. The nanoindentation hardness increased by a range of 4.11–6.31 GPa and 3.98–5.52 GPa with an improvement in the strain rate ranging from 0.001 to 1 s^{-1}. The alloy with only the linear scanning strategy and 90° rotation scanning strategy has a strain-rate sensitivity exponent m = 0.053 ± 0.014 and m = 0.047 ± 0.009, respectively.

The PBF-EB AM Ti6Al4V alloy with only the linear scanning strategy has better creep resistance properties than the alloy with a 90° rotation scanning strategy. Increasing peak holding load (10–100 mN) led to the dramatic increment of creep displacement (15–95 nm and 25–225 nm) and the creep behavior was mainly dominated by dislocation motion during deformation induced by the indenter.

Author Contributions: Conceptualization, H.P. and W.F.; validation, H.P. and Y.Y.; formal analysis, H.P., W.F., C.D. and S.H.; investigation, H.P., X.W., B.L. and S.H.; data curation, H.P. and X.W.; writing—original draft preparation, H.P. and W.F.; writing—review and editing, C.D. and Y.Y.; visualization, H.P., W.F. and B.L.; project administration, W.F. and Y.Y.; funding acquisition, W.F. and Y.Y. All authors have read and agreed to the published version of the manuscript.

Funding: This research was funded by Key Area R&D Program of Guangdong Province, China (No. 2018B090904004), Science and Technology Program of Guangzhou, China (No. 201907010010), Science and Technology Development Program provided by Guangdong Academy of Sciences, China (No. 2018GDASCX-0803), Guangzhou Economic and Technological Development Zone Program (No. 2019GH19).

Institutional Review Board Statement: Not applicable.

Informed Consent Statement: Not applicable.

Data Availability Statement: The data presented in this study are available on request from the corresponding author after obtaining permission of authorized person.

Conflicts of Interest: The authors declare that they have no known competing financial interests or personal relationships that could appear to influence the work reported in this paper.

References

1. Liu, G.; Zhang, X.F.; Chen, X.L.; He, Y.H.; Cheng, L.Z.; Huo, M.K.; Yin, J.N.; Hao, F.Q.; Chen, S.Y.; Wang, P.Y.; et al. Additive manufacturing of structural materials. *Mater. Sci. Eng.* **2021**, *9*, 100596. [CrossRef]
2. Cheng, B.; Price, S.; Lydon, J.; Cooper, K.; Chou, K. On Process Temperature in Powder-Bed Electron Beam Additive Manufacturing: Model Development and Validation. *ASME J. Manuf. Sci. Eng.* **2014**, *136*, 1–12. [CrossRef]
3. Bourell, D.; Kruth, J.P.; Leu, M.; Levy, G.; Rosen, D.; Beese, A.M.; Clare, A. Materials for additive manufacturing. *Ann. Manuf. Technol.* **2017**, *66*, 659–681. [CrossRef]
4. Zhao, Y.F.; Koizumi, Y.; Aoyagi, K.; Yamanaka, K.; Chiba, A. Isothermal γ → ε phase transformation behavior in a Co-Cr-Mo alloy depending on thermal history during electron beam powder-bed additive manufacturing. *J. Mater. Sci. Technol.* **2020**, *50*, 162–170. [CrossRef]

5. Gao, B.; Peng, H.; Liang, Y.; Lin, J.; Chen, B. Electron beam melted TiC/high Nb–TiAl nanocomposite: Microstructure and mechanical property. *Mater. Sci. Eng. A* **2021**, *811*, 141059. [CrossRef]
6. Kuwabara, K.; Shiratori, H.; Fujieda, T.; Yamanaka, K.; Koizumi, Y.; Chiba, A. Mechanical and corrosion properties of AlCoCrFeNi high-entropy alloy fabricated with selective electron beam melting. *Addit. Manuf.* **2018**, *23*, 264–271. [CrossRef]
7. Wang, P.; Nai, M.L.S.; Sin, W.J.; Lu, S.L.; Zhang, B.C.; Bai, J.M.; Song, J.; Wei, J. Realizing a full volume component by in-situ welding during electron beam melting process. *Addit. Manuf.* **2018**, *22*, 375–380. [CrossRef]
8. Klimov, A.S.; Bakeev, I.Y.; Dvilis, E.S.; Oks, E.M.; Zenin, A.A. Electron beam sintering of ceramics for additive manufacturing. *Vacuum* **2019**, *169*, 108933. [CrossRef]
9. Leon, A.; Levy, G.K.; Ron, T.; Shirizly, A.; Aghion, E. The effect of hot isostatic pressure on the corrosion performance of Ti-6Al-4V produced by an electron-beam melting additive manufacturing process. *Addit. Manuf.* **2020**, *33*, 101039.
10. Popov, V.V.; Muller-Kamskii, G.; Katz-Demyanetz, A.; Kovalevsky, A.; Usov, S.; Trofimcow, D.; Dzhenzhera, G.; Koptyug, A. Additive manufacturing to veterinary practice: Recovery of bony defects after the osteosarcoma resection in canines. *Biomed. Eng. Lett.* **2019**, *9*, 97–108. [CrossRef] [PubMed]
11. Levy, G.K.; Kafri, A.; Ventura, Y.; Leon, A.; Vago, R.; Goldman, J.; Aghion, E. Surface stabilization treatment enhances initial cell viability and adhesion for biodegradable zinc alloys. *Mater. Lett.* **2019**, *248*, 130–133. [CrossRef]
12. Popov, V.V.; Muller-Kamskii, G.; Kovalevsky, A.; Dzhenzhera, G.; Strokin, E.; Kolomiets, A.; Ramon, J. Design and 3D-printing of titanium bone implants: Brief review of approach and clinical cases. *Biomed. Eng. Lett.* **2018**, *8*, 337–344. [CrossRef]
13. Rahmani, R.; Antonov, M.; Brojan, M. Lightweight 3D printed Ti6Al4V-AlSi10Mg hybrid composite for impact resistance and armor piercing shielding. *J. Mater. Res. Technol.* **2020**, *9*, 13842–13854. [CrossRef]
14. Rahmani, R.; Brojan, M.; Antonov, M.; Prashanth, K.G. Perspectives of metal-diamond composites additive manufacturing using SLM-SPS and other techniques for increased wear-impact resistance. *Int. J. Refract. Met. Hard Mater.* **2020**, *88*, 105192. [CrossRef]
15. Wu, Y.C.; Kuo, C.N.; Chung, Y.C.; Ng, C.H.; Huang, J.C. Effects of electropolishing on mechanical properties and bio-corrosion of Ti6Al4V fabricated by electron beam melting additive manufacturing. *Materials* **2019**, *12*, 1466. [CrossRef]
16. Tiferet, E.; Ganor, M.; Zolotaryov, D.; Garkun, A.; Hadjadj, A.; Chonin, M.; Ganor, Y.; Noiman, D.; Halevy, I.; Tevet, O.; et al. Mapping the tray of electron beam melting of Ti-6Al-4V: Properties and microstructure. *Materials* **2019**, *12*, 1470. [CrossRef] [PubMed]
17. Radlof, W.; Benz, C.; Heyer, H.; Sander, M. Monotonic and fatigue behavior of EBM manufactured Ti-6Al-4V solid samples: Experimental, analytical and numerical investigations. *Materials* **2020**, *13*, 4642. [CrossRef]
18. Neikter, M.; Colliander, M.; Schwerz, C.d.A.; Hansson, T.; Akerfeldt, P.; Pederson, R.; Antti, M.L. Fatigue crack growth of electron beam melted Ti-6Al-4V in high-pressure hydrogen. *Materials* **2020**, *13*, 1287. [CrossRef]
19. Tan, X.P.; Chandra, S.; Kok, Y.; Tor, S.B.; Seet, G.; Loh, N.H.; Liu, E. Revealing competitive columnar grain growth behavior and periodic microstructural banding in additively manufactured Ti-6Al-4V parts by selective electron beam melting. *Materialia* **2019**, *7*, 100365. [CrossRef]
20. Wei, C.B.; Ma, X.L.; Yang, X.J.; Zhou, M.; Wang, C.M.; Zheng, Y.F.; Zhang, W.P.; Li, Z.J. Microstructural and property evolution of Ti6Al4V powders with the number of usage in additive manufacturing by electron beam melting. *Mater. Lett.* **2018**, *221*, 111–114. [CrossRef]
21. Oliveira, V.M.C.A.; Silva, M.C.L.; Pinto, C.G.; Suzuki, P.A.; Machado, J.P.B.; Chad, V.M.; Barboza, M.J.R. Short-term creep properties of Ti-6Al-4V alloy subjected to surface plasma carburizing process. *J. Mater. Res. Technol.* **2015**, *4*, 359–366. [CrossRef]
22. Kim, Y.K.; Youn, S.J.; Kim, S.W.; Hong, J.; Lee, K.A. High-temperature creep behavior of gamma Ti-48Al-2Cr-2Nb alloy additively manufactured by electron beam melting. *Mater. Sci. Eng. A* **2019**, *763*, 138176. [CrossRef]
23. Yavari, P.; Langdon, T.G. An examination of the breakdown in creep by viscous glide in solid solution alloys at high stress levels. *Acta Metall.* **1982**, *30*, 2181–2196. [CrossRef]
24. Gouldstone, A.; Koh, H.J.; Zeng, K.Y.; Giannakopoulos, A.E.; Suresh, S. Discrete and continuous deformation during nanoindentation of thin films. *Acta Mater.* **2000**, *48*, 2277–2295. [CrossRef]
25. Xu, Z.L.; Zhang, H.; Li, W.H.; Mao, A.Q.; Wang, L.; Song, G.S.; He, Y.Z. Microstructure and nanoindentation creep behavior of CoCrFeMnNi high-entropy alloy fabricated by selective laser melting. *Addit. Manuf.* **2019**, *28*, 766–771. [CrossRef]
26. Sun, S.; Gao, P.; Sun, G.X.; Cai, Z.; Hu, J.J.; Han, S.; Lian, J.S.; Liao, X.Z. Nanostructuring as a route to achieve ultra-strong high- and medium-entropy alloys with high creep resistance. *J. Alloys Compd.* **2020**, *830*, 154656. [CrossRef]
27. Jun, T.S.; Armstrong, D.E.J.; Britton, T.B. A nanoindentation investigation of local strain rate sensitivity in dual-phase Ti alloys. *J. Alloys Compd.* **2016**, *672*, 282–291. [CrossRef]
28. Li, H.; Ngan, A.H.W. Size effects of nanoindentation creep. *J. Mater. Res.* **2004**, *19*, 513–522. [CrossRef]
29. Wang, C.L.; Lai, Y.H.; Huang, J.C.; Nieh, T.G. Creep of nanocrystalline nickel: A direct comparison between uniaxial and nanoindentation creep. *Scr. Mater.* **2010**, *62*, 175–178. [CrossRef]
30. Goodall, R.; Clyne, T.W. A critical appraisal of the extraction of creep parameters from nanoindentation data obtained at room temperature. *Acta Mater.* **2006**, *54*, 5489–5499. [CrossRef]
31. Poisl, W.H.; Oliver, W.C.; Fabes, B.D. The relationship between indentation and uniaxial creep in amorphous selenium. *J. Mater. Res.* **1995**, *10*, 2024–2032. [CrossRef]
32. Choi, I.C.; Yoo, B.G.; Kim, Y.J.; Jang, J.I. Indentation creep revisited. *J. Mater. Res.* **2011**, *27*, 3–11. [CrossRef]

33. Shen, L.; Cheong, W.C.D.; Foo, Y.L.; Chen, Z. Nanoindentation creep of tin and aluminium: A comparative study between constant load and constant strain rate methods. *Mater. Sci. Eng. A* **2012**, *532*, 505–510. [CrossRef]
34. Song, J.; Wu, W.H.; Zhang, L.; He, B.B.; Lu, L.; Ni, X.Q.; Long, Q.L.; Zhu, G.L. Role of scanning strategy on residual stress distribution in Ti-6Al-4V alloy prepared by selective laser melting. *Optik* **2018**, *170*, 342–352. [CrossRef]
35. Wang, X.; Lv, F.; Shen, L.D.; Liang, H.X.; Xie, D.Q.; Tian, Z.J. Influence of island scanning strategy on microstructures and mechanical properties of direct laser-deposited Ti–6Al–4V structures. *Acta Metall. Sin.* **2018**, *32*, 1173–1180. [CrossRef]
36. Peng, H.L.; Hu, L.; Li, L.J.; Zhang, L.Y.; Zhang, X.L. Evolution of the microstructure and mechanical properties of powder metallurgical high-speed steel S390 after heat treatment. *J. Alloys Compd.* **2018**, *740*, 766–773. [CrossRef]
37. Kenel, C.; Grolimund, D.; Li, X.; Panepucci, E.; Samson, V.A.; Sanchez, D.F.; Marone, F.; Leinenbach, C. In situ investigation of phase transformations in Ti-6Al-4V under additive manufacturing conditions combining laser melting and high-speed micro-X-ray diffraction. *Sci. Rep.* **2017**, *7*, 16358. [CrossRef] [PubMed]
38. Zhao, X.L.; Li, S.J.; Zhang, M.; Liu, Y.D.; Sercombe, T.B.; Wang, S.G.; Hao, Y.L.; Yang, R.; Murr, L.E. Comparison of the microstructures and mechanical properties of Ti–6Al–4V fabricated by selective laser melting and electron beam melting. *Mater. Des.* **2016**, *95*, 21–31. [CrossRef]
39. Ma, R.X.; Liu, Z.Q.; Wang, W.B.; Xu, G.J.; Wang, W. Microstructures and mechanical properties of Ti6Al4V-Ti48Al2Cr2Nb alloys fabricated by laser melting deposition of powder mixtures. *Mater. Charact.* **2020**, *164*, 110321. [CrossRef]
40. Chen, G.L.; Xu, X.J.; Teng, Z.K.; Wang, Y.L.; Lin, J.P. Microsegregation in high Nb containing TiAl alloy ingots beyond laboratory scale. *Intermetallics* **2007**, *15*, 625–631. [CrossRef]
41. Takashima, T.; Koizumi, Y.; Li, Y.P.; Yamanaka, K.; Saito, T.; Chiba, A. Effect of Building Position on Phase Distribution in Co-Cr-Mo Alloy Additive Manufactured by Electron-Beam Melting. *Mater. Trans.* **2016**, *57*, 2041–2047. [CrossRef]
42. Peng, H.L.; Hu, L.; Li, L.J.; Gao, J.X.; Zhang, Q. On the correlation between $L1_2$ nanoparticles and mechanical properties of $(NiCo)_{52+2x}(AlTi)_{4+2x}Fe_{29-4x}Cr_{15}$ (x=0-4) high-entropy alloys. *J. Alloys Compd.* **2020**, *817*, 152750. [CrossRef]
43. VNguyen, L.; Kim, E.A.; Yun, J.; Choe, J.; Yang, D.Y.; Lee, H.S.; Lee, C.W.; Yu, J.H. Nano-mechanical Behavior of H13 Tool Steel Fabricated by a Selective Laser Melting Method. *Metall. Mater. Trans. A* **2019**, *50*, 523–528.
44. Duan, Z.; Pei, W.; Gong, X.; Chen, H. Superplasticity of Annealed H13 Steel. *Materials* **2017**, *10*, 870. [CrossRef] [PubMed]
45. Surand, L.B.; Ruau, J.; Viguier, B. Creep behavior of Ti-6Al-4V from 450 °C to 600 °C, U.P.B. *Sci. Bull.* **2014**, *76*, 185–196.
46. Barboza, M.J.R.; Perez, E.A.C.; Medeiros, M.M.; Reis, D.A.P.; Nono, M.C.A.; Neto, F.P.; Silva, C.R.M. Creep behavior of Ti–6Al–4V and a comparison with titanium matrix composites. *Mater. Sci. Eng. A* **2006**, *428*, 319–326. [CrossRef]

Article

Temperature Profile in Starch during Irradiation. Indirect Effects in Starch by Radiation-Induced Heating

Mirela Braşoveanu and Monica R. Nemţanu *

Electron Accelerators Laboratory, National Institute for Lasers, Plasma and Radiation Physics 409 Atomiştilor St., P.O. Box MG-36, 077125 Bucharest-Măgurele, Romania; mirela.brasoveanu@inflpr.ro
* Correspondence: monica.nemtanu@inflpr.ro

Abstract: Present research deals with exposure of granular starch to the accelerated electron of 5.5 MeV energy in order to examine: (i) the temperature evolution in starch within an irradiation process and (ii) the indirect effects generated in starch by radiation-induced heating. The temperature evolution in potato and corn starches within the irradiation process was investigated by placing two different sensors inside each starch batch and recording the temperature simultaneously. Each starch batch was sampled into distinct location sectors of different absorbed radiation levels. The output effects in each sample were analyzed through physicochemical properties such as moisture content, acidity and color attributes. The outcomes showed that a starch temperature profile had different major stages: (i) heating during irradiation, (ii) post-irradiation heating, up to the maximum temperature is reached, and (iii) cooling to the room temperature. A material constant with signification of a relaxation time was identified by modeling the temperature evolution. Changes of the investigated properties were induced both by irradiation and radiation-induced heating, depending on the starch type and the batch sectors. Changes in the irradiated batch sectors were explained by irradiation and radiation-induced heating whereas changes in the sector of non-irradiated starch were attributed only to the heating.

Keywords: electron beam; corn starch; potato starch; moisture content; specific heat capacity; pH; color parameters

Citation: Braşoveanu, M.; Nemţanu, M.R. Temperature Profile in Starch during Irradiation. Indirect Effects in Starch by Radiation-Induced Heating. *Materials* **2021**, *14*, 3061. https://doi.org/10.3390/ma14113061

Academic Editor: Katia Vutova

Received: 15 April 2021
Accepted: 31 May 2021
Published: 3 June 2021

Publisher's Note: MDPI stays neutral with regard to jurisdictional claims in published maps and institutional affiliations.

Copyright: © 2021 by the authors. Licensee MDPI, Basel, Switzerland. This article is an open access article distributed under the terms and conditions of the Creative Commons Attribution (CC BY) license (https://creativecommons.org/licenses/by/4.0/).

1. Introduction

Starch is one of the most widespread natural polymers, being composed of two different fractions: amylose (linear fraction) and amylopectin (branched fraction). Starch is used in various food and non-food applications such as pharmaceuticals and biomedical products, packaging materials, textiles, adhesives, etc. Native starch has limited functionality in technological applications due to its poor processing properties such as high viscosity, tendency for retrogradation, lack of thermal stability. However, it has a range of very significant advantages being a cheap, renewable, non-toxic, and widely available biodegradable raw material [1]. Thus, the need for starch modified by using emerging ecological techniques has gradually increased worldwide in recent years [2]. One of these techniques is based on ionizing radiation (gamma rays or electron beams) through which starch can be easily modified by degradation, crosslinking or grafting processes [3–6].

Electron beam irradiation is widely used in material processing, involving both chemical and thermal effects. The chemical effects due to ion and free radical generation [7,8] are connected with radiation chemistry. The thermal effects are especially related to the metal and alloy refining, melting and welding applications [9–11], where the transformation of the radiation energy into heat is done deliberately. It is to be remembered that the overlapping of thermal effects and radiation-induced chemical effects is avoided most of the times, especially for heat-sensitive materials. Among heat-sensitive materials can be found biological materials, which are sensitive to temperature variation, as well as materials that undergo phase transitions in temperature ranges slightly higher than the ambient temperature. As an example, granular starch undergoes an irreversible transition known as gelatinization in the presence of sufficient moisture above a certain temperature, generally between 60 and 80 °C [12]. In this context and in the absence of knowledge on this topic in literature, there is reason to believe that the heat transferred to starch from radiation field can make additional contribution to changes of the starch properties by irradiation. The water content of the starch can be diminished by evaporation due to the developed heat during irradiation. In this way, the reaction environment would be changed as well as the specific heat capacity of the starch. All these aspects actually underline the importance of knowing the dynamics of the temperature in starch during irradiation in order to control the final results of irradiation. In our previous work [13] on this topic, we reported a first theoretical approach concerning the temperature distribution in the granular corn starch exposed to electron beam irradiation through a semi-analytical model based on the heat equation in Cattaneo-Vernote formalism and solved by using the integral transform technique on finite domains. To the best of our knowledge, there are no other studies that explore this thematic area.

Furthermore, in the present work, we aimed out to show the evolution of temperature in granular starch subjected to an irradiation process. At the same time, the presence of indirect effects produced in starch by electron-beam induced heating was highlighted. For this purpose, starches from two different botanical sources, potato and corn, were exposed to an accelerated electron field in a selected irradiation setup (geometry and irradiation parameters). For both types of starch, the temperature was measured simultaneously at two points in the batch. The analysis of batches was performed by sampling them into three distinct location sectors and evaluating some physicochemical properties such as moisture content, pH and color attributes. The findings of this work are relevant for any research related to the irradiation of starch-based materials, especially whenever the research results are transferred at a large scale. An important advantage in the radiation application comes from the possibility of irradiating samples with large dimensions, but in this case the cooling process is more difficult and unforeseen thermal effects can occur.

2. Materials and Methods

2.1. Materials

Potato starch (S4251; moisture content: ~18%) and corn starch (S4126; moisture content: ~11%) used in the experiments were purchased from Sigma-Aldrich Company (St. Louis, MO, USA).

2.2. Electron Beam (E-Beam) Irradiation

The irradiation of starch powder in solid state was performed in static mode by using an e-beam generated by a linear accelerator ALID-7 of energy of 5.5 MeV (NILPRP, Bucharest-Măgurele, Romania). The electron accelerator facility is of traveling-wave type which uses microwaves in the S-band at 2.99 GHz propagating in a disk-loaded tube of about 2 m long, and the microwaves are produced by an EEV-M5125 type magnetron delivering 2 MW of power in pulses of 4 µs [14,15]. The ALID-7 accelerator is used for various applied radiation researches [16–18]. The e-beam had a Gaussian-like profile of dose rate distribution, which was canceled at a distance of r = 37 mm from the beam axis at a distance of 220 mm from accelerator window exit. The maximum dose rate of $\mathcal{D}_0$ = 12 kGy/min in the beam center, on the irradiation sample surface, was reached by using a mean beam intensity of 4 µA. The sample irradiation was carried out for a period of 435 s, at the room temperature (22 ± 1 °C) and ambient pressure in air.

2.3. Irradiation Geometry and Temperature Measurement

The starch powder was placed in a cylindrical cardboard box (Figure 1) so that the density was 540 ± 10 kg/m³ for both types of starch. Consequently, the maximum depth of electrons in starch determined according to the ISO/ASTM 51649:2002(E) [19] was P_{max} = 55 mm. The box had the radius R = 65 mm and the height H = 100 mm and its symmetry axis coincided with the e-beam axis during irradiation. Two temperature sensors (K-type thermocouple of a multifunctional digital multimeter DVM891, Velleman®, Gavere, Belgium) were placed inside the starch batch: S1 on the e-beam axis and S2 at $r = 37$ mm from the axis. Both sensors were located at a depth of 55 mm, which is equal to the maximum depth of the electrons in starch, in order to avoid that these sensors directly absorb heat/energy from the radiation field.

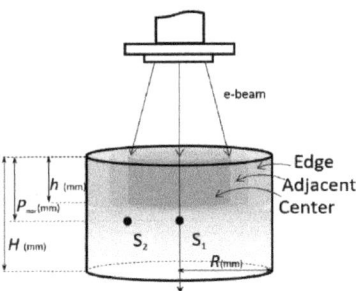

Figure 1. Experimental set-up.

The temperature values were simultaneously recorded by both sensors during starch irradiation and after irradiation, namely during the heating due to the radiation field as well as during the cooling toward room temperature.

2.4. Sampling of Batches

Each batch of irradiated starch was divided into three distinct sectors, concentric around the symmetry axis. Each starch sample was collected from each location sector to a depth of $h = 40$ mm (Figure 1). The collected samples were then investigated separately.

The sector named Center, located 37 mm around the center and exposed to the proper absorbed dose rate and the highest heating, was associated with the temperature recorded by the sensor S1. Thus, $S1_{potato}$ and $S1_{corn}$ refer to the temperature in this sensor for potato starch sample and corn starch sample, respectively.

The second sector, named Adjacent, was located around the Center, up to 50 mm from the beam symmetry axis, being associated with the temperature recorded by the sensor S2. Thus, $S2_{potato}$ and $S2_{corn}$ refer to the temperature in this sensor for potato starch sample and corn starch sample, respectively. A "residual" absorbed dose rate in the Adjacent sector was estimated to be less than 7% from the central maximum dose rate due to the secondary electrons.

The Edge sector was located between 50 and 65 mm from the center, at the periphery of the sample, free of radiation, being associated with a small variation of temperature, but not recorded.

2.5. Moisture Content Determination

The moisture content of the samples was measured by using an IR-200 moisture analyzer (Denver Instruments Company, Denver, CO, USA) at 105 °C for 45 min.

2.6. pH Measurement

Starch samples were mixed in distilled water by magnetic stirring on bath water at 85 °C for 30 min, and the mixtures of 1% (w/v) concentration were then allowed to stand at room temperature (25 ± 1 °C). After cooling, pH was measured at 25 ± 1 °C with an inoLab® 7110 pH-meter (WTW, Weilheim, Germany).

2.7. Optical Measurements

Starch samples were mixed in distilled water by continuously magnetic stirring on bath water at 85 °C for 30 min, and the mixtures of 1% (w/v) concentration were then allowed to stand at room temperature (25 ± 1 °C). After cooling, the UV-Vis spectra of the samples were recorded in a Cary 100 Bio spectrophotometer (Varian, Inc., Walnut Creek, CA, USA). Colorimetric features of the samples were measured in absorbance in the visible region (360–830 nm), for standard illuminant D65 (daylight source), observer angle of 10° (perception angle of a human observer). The colorimetric attributes expressed as CIELCH parameters (L^*, C^*, $h°$) were analyzed by using the Color Application of Cary Win UV v. 3.10 software (Vary, Inc., Walnut Creek, CA, USA, 2006).

2.8. Statistical Approach

The results reported are expressed by means of values ± standard deviation of triplicate determinations. The processing of experimental data was performed using OriginPro 8.1 (OriginLab Corporation, Northampton, MA, USA, 2016), Microsoft® Excel 2010 (Microsoft Corporation, Redmond, WA, USA), and InfoStat versión 2018 [20]. The data were analyzed by using analysis of variance with Fisher LSD (least significant differences) post-hoc test to discern the statistical difference. A probability value $p \leq 0.05$ was considered as statistically significant.

3. Results and Discussion

3.1. Temperature Profile

The evolution of the temperature recorded by the two sensors, for each type of starch, during irradiation and cooling is presented in Figure 2. It was observed that there are three major stages: (i) heating during irradiation, up to t = 435 s, (ii) post-irradiation heating, up to the maximum temperature is reached, and (iii) cooling to the room temperature. This finding is consistent with the fact that the heat continues to propagate in starch for a while, in connection with the place where the sensors are placed, after the irradiation process is stopped and before the cooling process begins [13]. The maximum temperature and the moment of its reaching differ for the two types of investigated starch, both in the first sensor, $S1_{potato}$ and $S1_{corn}$, but even more visible in the second sensor, $S2_{potato}$ and $S2_{corn}$. This differentiation of behavior can be attributed to the different thermal properties of these two types of investigated starch.

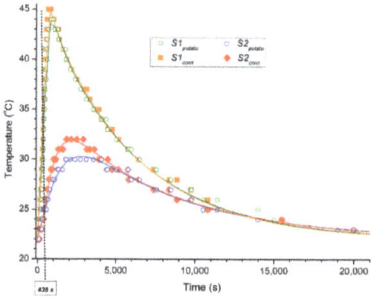

Figure 2. Temperature profiles through sensors S1 and S2 for both irradiated potato and corn starches.

During the irradiation stage, the potato starch had a lower heating rate than corn starch, reaching a temperature of 36 °C in $S1_{potato}$ compared to 38 °C in $S1_{corn}$ after 435 s. Also, the potato starch continued to heat up post-irradiation for a longer time than corn starch, visible in both sensors. Therefore, the potato starch needed 364 s to reach the maximum temperature in S1 while the corn starch needed only 216 s. Conversely, the maximum temperature was higher for corn starch compared to the potato starch, with 1 °C in S1, and with 2 °C in S2. However, the time to reach the peak temperature in S2 appeared to be comparable for the two starches, but this maximum was maintained much longer for potato starch than corn starch. In this way, the last stage started later with lower values for potato starch than corn starch, in both sensors. The cooling curves showed an exponential trend, so that the temperatures for potato starch and corn starch in the first sensor, $S1_{potato}$ and $S1_{corn}$, respectively, overlapped quite well, especially after about 2000 s. Also, in the second sensor, the cooling process was similar for the two starches after about 5000 s.

By analyzing the cooling curves, it can be noticed that their evolution followed an exponential law, in each sensor:

$$T = T_0 e^{-t/\tau_0} + 22, \qquad (1)$$

where T is the temperature in each sensor during cooling process (°C), T_0 is a temperature parameter (°C), which depends on both the maximum temperature and the time considered as the beginning of the cooling process, t is the time (s), τ_0 is a parameter that has the signification of a relaxation time (s), namely the time required for starch to decrease the temperature to the equilibrium temperature e times.

The values of τ_0 obtained by fitting the experimental data of the cooling process are displayed for each starch and each sensor in the first row of Table 1. The parameter τ_0 estimated in this way can characterize the starch thermal behavior, but also it depends on the geometry characterizing the starch at the time of temperature measuring.

Table 1. Values of the parameters in the Equations (1) and (2) as result of the experimental data fitting.

Parameter	Potato Starch		Corn Starch	
	$S1_{potato}$	$S2_{potato}$	$S1_{corn}$	$S2_{corn}$
τ_0 (s)	6260 ± 108	7585 ± 409	5981 ± 116	6964 ± 248
τ_i (s)	381 ± 31	1914 ± 448	267 ± 22	1117 ± 252
$\tau_0 - \tau_i$ (s)	5879 ± 139	5671 ± 857	5714 ± 138	5847 ± 500
Δt (s)	670 ± 19	3993 ± 1675	627 ± 15	3834 ± 2126
T_0 (°C)	23.0 ± 0.3	13.1 ± 0.7	24.6 ± 0.2	14.7 ± 0.3
T_i (°C)	40.9 ± 3.6	6.9 ± 4.0	45.5 ± 4.9	5.7 ± 4.1
R^2	0.9944	0.9754	0.9973	0.9866

On the other hand, we were able to find a function that also describes the evolution of the starch temperature for post-irradiation heating process:

$$T = \left[T_i \frac{t - \Delta t}{\tau_i} e^{-t/\tau_i} + T_0 \right] e^{-t/\tau_0} + 22, \qquad (2)$$

where T is the temperature in each sensor after irradiation stopping (°C), T_i (°C), τ_i (s) and Δt (s) are parameters that characterize the post-irradiation heating, t is the time (s), T_0 is a temperature parameter, associated with the temperature at which the cooling started, τ_0 is the parameter with significance of relaxation time (s).

Keeping the values of τ_0 found for the cooling process, we determined the other parameters that best fit (R^2) the experimental data for the post-irradiation heating processes, and the values obtained are shown in Table 1. By carefully evaluating the determined parameter values, we noticed that T_0 is the difference from the room temperature to the maximum value reached in each sensor, and τ_i can express the time required to reach the maximum temperature after irradiation in each sensor. T_i and Δt take values that describe the temperatures in sensors when the irradiation is off as well as the shape of the curves around the maximum value. The parameters T_i, τ_i, and Δt depend more on the irradiation geometry (S1 vs. S2) than on the type of starch ($S1_{potato}$ vs. $S1_{corn}$ and $S2_{potato}$ vs. $S2_{corn}$, respectively). Note that even though τ_0 and τ_i depend on the starch type and geometry, the values of their difference $\tau_0 - \tau_i$ are not significantly different. This finding suggests that $\tau_0 - \tau_i$ could be a material constant, and the starch relaxation time could actually have values around 5800 s.

3.2. Moisture Content and Specific Heat Capacity

The moisture content for the starch sampled from the three considered locations (Center, Adjacent and Edge) was evaluated for each type of starch (potato and corn) together with the corresponding control samples. The results presented in Table 2 show that the moisture content of both types of starches is insignificantly ($p > 0.05$) affected in the Edge compared to that of the control samples. A decrease in moisture content was observed in the Center and Adjacent sectors compared to the control samples for both starches. However, corn starch showed a drastic decrease ($p \leq 0.05$) of moisture content only in the Center. At the same time, the potato starch, with a higher moisture content, suffered ($p \leq 0.05$) the highest loss, of over 20% in the *Center*. It is noteworthy that potato starch showed a higher moisture loss than corn starch for each sector, which can be explained by the origin of different botanical sources and, implicitly, their different structural arrangement.

Table 2. Values of the moisture content ($u\%$) and specific heat capacity (c_p) for studied starches.

Location Sector	Potato Starch		Corn Starch	
	u (%)	c_p (J kg^{-1} K^{-1})	u (%)	c_p (J g^{-1} K^{-1})
Control sample (native)	17.7 ± 0.5 [a]	1434 ±15 [a]	11.2 ± 0.3 [a]	1215 ± 10 [a]
Edge	17.5 ± 0.1 [ab]	1429 ± 5 [ab]	11.2 ± 0.4 [a]	1216 ± 12 [a]
Adjacent	16.5 ± 0.4 [b]	1395 ± 14 [b]	10.6 ± 0.4 [a]	1196 ± 12 [a]
Center	13.9 ± 0.4 [c]	1308 ± 14 [c]	9.1 ± 0.5 [b]	1147 ± 17 [b]

Values within each column with different superscripts are significantly different ($p \leq 0.05$).

The starch dehydration could be explained by a water vapor transport that accompanied the cooling process which occurred by heat dissipation from the Center to the Edge. Moisture content losses in the Center and Adjacent sectors are in accordance with the estimated values for the temperature parameter T_0. Thus, these values are associated with the maximum temperature reached in the sensors assigned to these location sectors. The different temperatures measured in a sensor for investigated samples could be explained by their different specific heat capacities, c_p. It is well known that the specific heat capacity c_p of a sample is directly related to its moisture content. Considering that $c_p = 3.35\,u + 0.84$ (kJ kg^{-1} K^{-1}) [21], where $u = u\%/100$ is the mass ratio of water in starch (kg kg^{-1}), the specific heat capacity values c_p were determined as shown in Table 2. Since moisture content was diminished due to irradiation, the specific heat capacity and, consequently, the thermal behavior of the samples were affected during irradiation and post-irradiation.

3.3. pH Evaluation

The native starches had an acidic character with a pH around 5.5 for both starches. Both starches showed the reduction ($p \leq 0.05$) of pH values for all investigated location sectors in comparison to the control sample (Table 3). Although both types of starches had a similar pH value decreasing in the Center, the corn starch suffered a greater reduction in pH value compared to potato starch for the Adjacent. The decreasing pH by e-beam irradiation in the presence of oxygen could be assigned to the fragmentation of starch molecules due to the free radicals and oxidative reactions leading to the formation of compounds with acidic chemical groups [22,23].

Table 3. pH values for investigated starch samples.

Location Sector	Potato Starch	Corn Starch
Control sample (native)	5.56 ± 0.06 [a]	5.43 ± 0.09 [a]
Edge	5.14 ± 0.18 [b]	5.32 ± 0.12 [a]
Adjacent	4.41 ± 0.19 [c]	4.12 ± 0.05 [b]
Center	3.57 ± 0.04 [d]	3.54 ± 0.03 [c]

Values within each column with different superscripts are significantly different ($p \leq 0.05$).

On the other hand, only potato starch had a significant reduction in pH value in the Edge in comparison to control sample. However, it should be noted here that the Edge sector was considered free of radiation, but with a small temperature variation, which led us to believe that the difference ($p \leq 0.05$) in pH value for potato starch can be totally attributed to a thermal phenomenon, even if it was a small one. This different behavior of the two types of starch can be attributed to their different amylose content. It is known that the thermal properties of starches from different botanical sources are influenced not only by the structural features, but the amylose content or amylopectin/amylose ratio, which leads to higher or less thermal stability of starch polymers [24,25]. There are reports showing that thermal stability decreased with increasing amylose content for both corn starch [26] and potato starch [27]. As the potato starch has a lower amylose content than the corn starch [28,29] and therefore it may show a greater thermal sensitivity, the change ($p \leq 0.05$) in its pH value due to the heating in the Edge can be explained.

3.4. Color Parameters

In the CIE $L^*C^*h°$ coordinate system, L^* is lightness (0 (black) → 100 (white)), chroma (saturation) C^* is the quantitative component of the color while hue $h°$ is the qualitative component of color, being expressed in degrees: 0° (red), 90° (yellow), 180° (green), 270° (blue). In these terms, native potato starch showed higher values of L^* and C^* than native corn starch. In other words, the potato starch had greater transparency and saturation in comparison to corn starch that showed the tendency of its color to grey. Instead, the hue $h°$ had similar values indicating the same slightly yellowish color hue for both native starches.

In our experiment, it was observed that the samples from the considered location sectors, for both starches, showed modified values ($p \leq 0.05$) for the parameters that characterize the color space $L^*C^*h°$ compared to the corresponding native counterparts. However, the corn starch from Edge showed values of color parameters similar to those of its native counterpart.

Although the color parameter evolution in opposite directions was noted for both potato and corn starches in the Adjacent and Center, all samples of both starches showed an intensification in terms of lightness by irradiation (Figure 3). Similar observations were previously reported by Nemțanu & Brașoveanu [30].

In the case of potato starch (Figure 3a), the chroma C^* decreased significantly while the hue $h°$ tended to green by moving from the red-yellow quadrant (0–90°) to the yellow-green one (90–180°) for the sample from the Center. In contrast, for corn starch (Figure 3b), the color saturation C^* of the sample from the Center intensified dramatically, and the hue $h°$ practically had only a slight shift to red (~2.5° for the sample from the Center), indicating the red-yellowish color trend.

In this study of the chromatic attributes, as in the case of pH, it was noticed that the potato starch from Edge showed significant changes compared to its native counterpart. Therefore, once again, it was found that potato starch, unlike corn starch, was affected by a thermal phenomenon induced by irradiation in a location sector (Edge) considered free of radiation.

Based on all these results and observations, it can be stated that radiation-induced heating can affect the physicochemical properties of starches, depending on their botanical source, moisture content, amylopectin/amylose content, and thermal sensitivity.

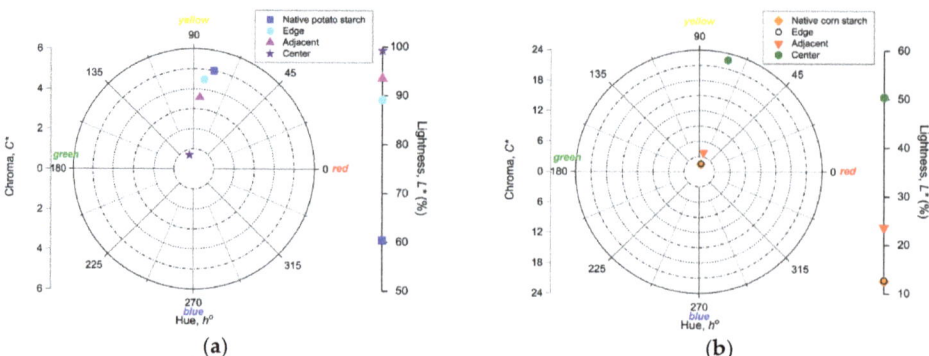

Figure 3. CIE $L^*C^*h^\circ$ parameters of the investigated samples: (**a**) potato starch and (**b**) corn starch.

4. Conclusions

The temperature evolution in two different granular starches (potato and corn) within an electron beam irradiation process was investigated. Also, the indirect effects that occurred in starch by radiation-induced heating were studied by sampling each starch batch into distinct location sectors having different absorbed radiation levels (maximum dose rate, "residual" dose rate, no radiation). The main findings of this study are:

1. The temperature profiles recorded in starch samples showed three major stages: (i) heating during irradiation, (ii) post-irradiation heating, up to the maximum temperature is reached, and (iii) cooling to the room temperature.
2. The maximum values of temperatures reached and the heating rates were different for both starches in the studied location sectors. However, the cooling rates were similar for both starches in each location sector thermally studied.
3. The evolution of the measured temperatures was modeled as a function of the parameters that depend on the starch type and the location sector considered inside the batch. Based on this modeling, a material constant having the significance of a relaxation time was identified with values around 5800 s.
4. Dehydration and changes in the values of the specific heat capacity, pH and color parameters of the starch were noticed due to the irradiation and radiation-induced heating, depending on the starch type and the batch sectors. The changes in the irradiated batch sectors (with maximum or "residual" dose rate) could be explained by irradiation and radiation-induced heating. On the other hand, the changes in the sector where the starch was practically not irradiated could be attributed only to the heating. Although the reached temperatures here were lower than in the other batch sectors, these changes cannot be ignored.

Therefore, for a heat-sensitive material like starch, specific experimental precautions are required to prevent the temperature rise and dehydration due to irradiation whenever indirect consequences are not desired. Moisture-containing materials may undergo a dehydration process by irradiation as a result of the transport of water vapors. At the same time, some thermal characteristics such as specific heat capacity can modify considerably due to the very high value of the water heat capacity. Our findings are of practical use for any experimental design in which the radiation-induced direct and indirect effects in the moisture-containing materials should be taken into consideration. When simulating processes due to irradiation, such as Monte Carlo method, it is necessary to take into account such variations in temperature and moisture content that can occur in a material. Thus, the main challenge in this research field is the enhanced control and mitigation of the radiation-induced thermal effects in order to enable the desired chemical effects on starch-based materials. Future work in the field should be performed to differentiate the effects induced in starch directly by irradiation from those developed indirectly by radiation-induced heating.

Author Contributions: Conceptualization, M.B. and M.R.N.; methodology, M.B. and M.R.N.; formal analysis, M.B. and M.R.N.; investigation, M.B. and M.R.N.; writing—original draft preparation, M.B.; writing—review and editing, M.B. and M.R.N.; supervision, M.R.N. All authors have read and agreed to the published version of the manuscript.

Funding: This research was funded by Romanian Ministry of Education and Research, under (1) Romanian National Nucleu Program LAPLAS VI—contract no. 16N/2019 and (2) National Interest Facilities of the National Institute for Lasers, Plasma and Radiation Physics—Electron Accelerators Laboratory.

Institutional Review Board Statement: Not applicable.

Informed Consent Statement: Not applicable.

Data Availability Statement: Data sharing is not applicable for this article.

Conflicts of Interest: The authors declare no conflict of interest. The funders had no role in the design of the study; in the collection, analyses, or interpretation of data; in the writing of the manuscript, or in the decision to publish the results.

References

1. Zarski, A.; Bajer, K.; Kapuśniak, J. Review of the most important methods of improving the processing properties of starch toward non-food applications. *Polymers* **2021**, *13*, 832. [CrossRef]
2. Mi, G.; Wang, T.; Li, J.; Li, X.; Xie, J. Phase separation affects the rheological properties of starch dough fortified with fish actomyosin. *RSC Adv.* **2021**, *11*, 9303–9314. [CrossRef]
3. Nemțanu, M.R.; Brașoveanu, M. Degradation of amylose by ionizing radiation processing. *Starch-Stärke* **2017**, *69*, 1600027. [CrossRef]
4. Ghobashy, M.M.; Abd El-Wahab, H.; Ismail, M.A.; Naser, A.M.; Abdelhai, F.; El-Damhougy, B.K.; Nady, N.; Meganid, A.S.; Alkhursani, S.A. Characterization of starch-based three components of gamma-ray cross-linked hydrogels to be used as a soil conditioner. *Mater. Sci. Eng. B* **2020**, *260*, 114645. [CrossRef]
5. López, O.V.; Ninago, M.D.; Lencina, M.M.S.; Ciolino, A.E.; Villar, M.A.; Andreucetti, N.A. Starch/Poly(ε-caprolactone) Graft Copolymers Synthetized by γ-Radiation and Their Application as Compatibilizer in Polymer Blends. *J. Polym. Environ.* **2019**, *27*, 2906–2914. [CrossRef]
6. Nemțanu, M.R.; Brașoveanu, M. Ionizing irradiation grafting of natural polymers having applications in wastewater treatment. In *Polymer Science: Research Advances, Practical Applications and Educational Aspects*; Méndez-Vilas, A., Solano-Martín, A., Eds.; Formatex Research Center: Badajoz, Spain, 2016; pp. 270–277.
7. Rummeli, M.H.; Ta, H.Q.; Mendes, R.G.; Gonzalez-Martinez, I.G.; Zhao, L.; Gao, J.; Fu, L.; Gemming, T.; Bachmatiuk, A.; Liu, Z. New frontiers in electron beam-driven chemistry in and around graphene. *Adv. Mater.* **2019**, *31*, 1800715. [CrossRef]
8. Siwek, M.; Edgecock, T. Application of electron beam water radiolysis for sewage sludge treatment—A review. *Environ. Sci. Pollut. Res.* **2020**, *27*, 42424–42448. [CrossRef]
9. Jokisch, T.; Doynov, N.; Ossenbrink, R.; Vesselin, G.M. Heat source model for electron beam welding of nickel-based superalloys. *Mater. Test.* **2021**, *63*, 17–28. [CrossRef]
10. Vutova, K.; Vassileva, V.; Stefanova, V.; Amalnerkar, D.; Tanaka, T. Effect of electron beam method on processing of titanium technogenic material. *Metals* **2019**, *9*, 683. [CrossRef]
11. Vutova, K.; Vassileva, V.; Koleva, E.; Stefanova, V.; Amalnerkar, D.P. Effects of process parameters on electron beam melting technogenic materials for obtaining rare metals. *J. Phys. Conf. Ser.* **2018**, *1089*, 012013. [CrossRef]
12. Copeland, L.; Blazek, J.; Salman, H.; Tang, M.C. Form and functionality of starch. *Food Hydrocoll.* **2009**, *23*, 1527–1534. [CrossRef]
13. Brașoveanu, M.; Oane, M.; Nemțanu, M.R. Heat transport in starch exposed to ionizing radiation: Experiment versus theoretical computer modeling. *Starch-Stärke* **2019**, *71*, 1900147. [CrossRef]
14. Martin, D.; Ighigeanu, D.; Toma, M.; Oproiu, C. Constructive and functional peculiarities of electron linear accelerators used in conjunction with microwave sources. In *Practical Aspects and Applications of Electron Beam Irradiation*; Nemțanu, M.R., Brașoveanu, M., Eds.; Research Signpost/Transworld Research Network: Trivandrum, India, 2011; pp. 1–16.
15. Ticoș, D.; Scurtu, A.; Oane, M.; Diplașu, C.; Giubega, G.; Călina, I.; Ticoș, C.M. Complementary dosimetry for a 6MeV electron beam. *Results Phys.* **2019**, *14*, 102377. [CrossRef]
16. Craciun, G.; Manaila, E.; Stelescu, M.D. New elastomeric materials based on natural rubber obtained by electron beam irradiation for food and pharmaceutical use. *Materials* **2016**, *9*, 999. [CrossRef]

17. Călina, I.; Demeter, M.; Scărișoreanu, A.; Sătulu, V.; Mitu, B. One step e-beam radiation cross-linking of quaternary hydrogels dressings based on chitosan-poly(vinyl-pyrrolidone)-poly(ethylene glycol)-poly(acrylic acid). *Int. J. Mol. Sci.* **2020**, *21*, 9236. [CrossRef] [PubMed]
18. Brașoveanu, M.; Nemțanu, M.R. Pasting properties modeling and comparative analysis of starch exposed to ionizing radiation. *Radiat. Phys. Chem.* **2020**, *168*, 108492. [CrossRef]
19. ASTM International. ISO/ASTM 51649:2002(E): Standard practice for dosimetry in an electron beam facility for radiation processing at energies between 300 keV and 25 MeV. In *Standards on Dosimetry for Radiation Processing*, 2nd ed.; ASTM International: West Conshohocken, PA, USA, 2004; pp. 133–152.
20. Di Rienzo, J.A.; Casanoves, F.; Balzarini, M.G.; Gonzalez, L.; Tablada, M.; Robledo, C.W. *InfoStat Versión 2018*; InfoStat Group, Facultad de Ciencias Agropecuarias, Universidad Nacional de Córdoba: Córdoba, Argentina, 2018.
21. Wilhelm, L.R.; Suter, D.A.; Brusewitz, G.H. *Food & Process Engineering Technology*; ASAE: St. Joseph, MI, USA, 2004; pp. 23–49.
22. Ershov, B.G. Radiation-chemical degradation of cellulose and other polysaccharides. *Russ. Chem. Rev.* **1998**, *67*, 315–334. [CrossRef]
23. Nemțanu, M.R.; Brașoveanu, M. Exposure of starch to combined physical treatments based on corona electrical discharges and ionizing radiation. Impact on physicochemical properties. *Radiat. Phys. Chem.* **2021**, *184*, 109480. [CrossRef]
24. Liu, X.; Wang, Y.; Yu, L.; Tong, Z.; Chen, L.; Liu, H.; Li, X. Thermal degradation and stability of starch under different processing conditions. *Starch-Stärke* **2012**, *65*, 48–60. [CrossRef]
25. Zhu, X.; He, Q.; Hu, Y.; Huang, R.; Shao, N.; Gao, Y. A comparative study of structure, thermal degradation, and combustion behavior of starch from different plant sources. *J. Therm. Anal. Calorim.* **2018**, *132*, 927–935. [CrossRef]
26. Liu, X.; Yu, L.; Xie, F.; Li, M.; Chen, L.; Li, X. Kinetics and mechanism of thermal decomposition of cornstarches with different amylose/amylopectin ratios. *Starch-Stärke* **2010**, *62*, 139–146. [CrossRef]
27. Stawski, D. New determination method of amylose content in potato starch. *Food Chem.* **2008**, *110*, 777–781. [CrossRef]
28. Wu, H.; Corke, H. Genetic variation in thermal properties and gel texture of *Amaranthus* starch. In *Cassava, Starch and Starch Derivatives, Proceedings of the International Symposium, Nanning, China, 11–15 November 1996*; Howeler, R.H., Oates, C.G., O'Brien, G.M., Eds.; Centro Internacional de Agricultura Tropica (CIAT): Cali-Palmira, Colombia, 1996; pp. 292–301.
29. Giuberti, G.; Gallo, A.; Moschini, M.; Masoero, F. In vitro production of short-chain fatty acids from resistant starch by pig faecal inoculum. *Animal* **2013**, *7*, 1446–1453. [CrossRef] [PubMed]
30. Nemțanu, M.R.; Brașoveanu, M. Functional properties of some non-conventional treated starches. In *Biopolymers*; Eknashar, M., Ed.; Scyio: Rijeka, Croatia, 2010; pp. 319–344.

Article

Behaviour of Impurities during Electron Beam Melting of Copper Technogenic Material

Katia Vutova [1,*], Vladislava Stefanova [2], Vania Vassileva [1] and Milen Kadiyski [3]

1. Institute of Electronics, Bulgarian Academy of Sciences, 1784 Sofia, Bulgaria; vvvania@abv.bg
2. Department of Metallurgy of Non-Ferrous Metals and Semiconductors Technologies, University of Chemical Technology and Metallurgy, 1756 Sofia, Bulgaria; vps@uctm.edu
3. Aurubis Bulgaria AD, Industrial Zone, 2070 Pirdop, Bulgaria; m.kadiyski@aurubis.com
* Correspondence: katia@van-computers.com

Abstract: The current study presents the electron beam melting (EBM) efficiency of copper technogenic material with high impurity content (Se, Te, Pb, Bi, Sn, As, Sb, Zn, Ni, Ag, etc.) by means of thermodynamic analysis and experimental tests. On the basis of the calculated values of Gibbs free energy and the physical state of the impurity (liquid and gaseous), a thermodynamic assessment of the possible chemical interactions occurring in the $Cu-Cu_2O-Me_x$ system in vacuum in the temperature range 1460–1800 K was made. The impact of the kinetic parameters (temperature and refining time) on the behaviour and the degree of removal of impurities was evaluated. Chemical and metallographic analysis of the obtained ingots is also discussed.

Keywords: copper technogenic material; electron beam; thermodynamic analysis; removal efficiency

1. Introduction

Copper continues to be one of the most important metals that are at the basis of the economic development of society and the human efforts to achieve a higher standard of living [1]. This is due to its unique physical, mechanical and chemical properties. At present, approximately 50% of the copper in Europe is produced by recycling. Copper recycling is becoming one of the main methods of producing copper, as it requires up to 85% less energy compared to the conventional production schemes [2].

The impurities content in secondary copper raw materials varies from 0.8% to 1.5% [3]. They usually contain a considerable amount of dissolved oxygen and sulphur and metal impurities such as lead, tin, iron, nickel, arsenic, antimony, zinc, bismuth, selenium, tellurium, gold, silver and others [3].

According to [4], copper-soluble impurities (such as Al, Sn, Zn, etc.) increase the mechanical properties but significantly reduce the electrical and thermal conductivity. Insoluble impurities (such as Pb, Bi) form eutectics which melt at lower temperatures, which worsen the hot treatment of copper under pressure. Non-metallic impurities (such as S, O) form eutectics which melt at higher temperatures that are separated at the boundaries of copper grains. This in turn leads to the appearance of brittleness of copper. Impurities that form intermetallic compounds (such as Se, Te) are separated as intermediate phases at the grain boundaries, leading to brittleness.

In traditional metallurgy, the process of refining Cu scrap is carried out by either pyrometallurgical or hydrometallurgical method [5]. In the pyrometallurgical scheme, the removal of impurities is done in anode furnaces and by electrolysis. According to the authors, about 85% of anode copper is subjected to electrolytic refining [5]. In hydrometallurgical schemes, the copper scrap is first dissolved in acids and then recovered, for example, by liquid phase extraction [6,7].

The requirements for the quality of copper, its chemical composition and structure are constantly increasing with the development of new branches of energy and electronics [8]. This necessitates the search for new, effective methods of melting.

Methods such as laser, plasma and electron beam melting (EBM) are successfully applied following the development of modern metallurgy and effective methods for refining metals and alloys [9–11]. Of these methods, the EBM method is particularly appropriate as it combines the advantages of vacuum and high-energy special electrometallurgy [12–16]. Under vacuum conditions, some reactions take place that are impossible at atmospheric pressure. The electron beam (heat source) and the high vacuum ensure degassing and a high degree of purification of the material, as well as uniformity of the chemical composition and homogeneous structure of the obtained ingots. Lack of additional impurities originating from the used water-cooled copper crucible and the ability to control the energy of the heat source regardless of the material and the size of the feedstock are additional important benefits of the method.

There are a number of publications in the literature related to the thermodynamics and kinetics of copper refining from impurities at atmospheric pressure [5,17–19]. Results of the modified pyrometallurgical processing of waste printed circuit boards were presented in [18], while pyrometallurgical refining of copper in an anode furnace was studied in [5] in order to improve the process. The behavior of tin and antimony was experimentally studied in secondary copper pyrometallurgical smelting conditions [19]. According to [17], depending on the degree of removal of impurities in the anodic refining, they can be divided into three groups: Group I–impurities separated relatively easily and completely (such as Fe, Co, Pb, Sn, S); Group II–impurities partially separated (such as As, Sb, Ni); Group III–impurities (such as Ag, Au, Se, Te, Bi) separated to a negligible extent.

This work is a continuation of the research conducted so far by the team [20]. In [20], the investigated material contained 99.83% copper and the influence of the beam power and duration of melting process on the purity of the refined material was studied, the electron affinity to oxygen of the investigated impurities and the vapor pressures of metallic impurities and their oxides were evaluated as well. The aim of this work is to study the behaviour of impurities (metallic and non-metallic) and the influence of thermodynamic and kinetic technological parameters (temperature of thermal treatment, refining time) on the refining efficiency and the structure of the obtained copper during EBM of copper technogenic material (99.45%). On the basis of the calculated values of Gibbs free energy under EBM conditions, a thermodynamic assessment of the possible chemical interactions occurring during the melting and refining of copper technogenic material with high impurity content (Se, Te, Pb, Bi, Sn, As, Sb, Zn, Ni, Ag, O, etc.) was made and the physical state of the base metal and metallic impurities was taken into account.

2. Materials and Methods

The experiments for copper technogenic material melting were performed using EBM installation with power 60 kW (ELIT-60) at the Physical problems of the EB technologies laboratory of the Institute of electronics, Bulgarian Academy of Sciences. ELIT-60 (Leybold GmbH, Cologne, Germany) is equipped with one electron gun (accelerating voltage of 24 kV), a feeding mechanism for horizontal input of the raw material, an extraction system (pulling mechanism), a water-cooled copper cylindrical crucible (a diameter of 50 mm) with moving bottom, where the molten metal solidifies and a circulation water cooling system–Figure 1. The operation vacuum pressure in the melting chamber is 3–6×10^{-3} Pa.

The copper content in the investigated technogenic material is 99.45% Cu (anode residues after an electrolysis process). The impurities with higher concentrations are: O (0.2251%), Ni (0.1%), As (0.07%), Se (0.0382%), Pb (0.021%), Bi (0.015%) and Ag (0.014%). With lower content are: S (0.0028%); Sb (0.0095%), Te (0.0068%), Sn (0.0034%) and Zn (0.0023%). The lowest is the content of Co, Cd, Fe, Au, etc. (<20 ppm), therefore these impurities are not taken into consideration in the further analysis.

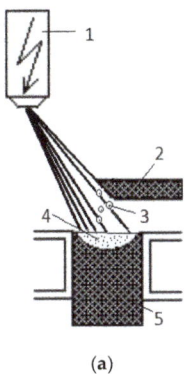 

(a)   (b)

Figure 1. (a) Principal scheme of the EBMR process: (1) electron optical system; (2) started metal rod; (3) generated droplets; (4) molten pool in the water-cooled crucible; (5) metal ingot; (b) fabricated copper sample.

The raw (initial) material was melted under single processing at melting powers of 6 kW (T = 1460 K), 7 kW (T = 1500 K), 13 kW (T = 1700 K) and 19.5 kW (T = 1800 K). At T = 1500 K the lengths of melting time are 15 min and 35 min, while at T = 1700 K the retention time is 20 min and 45 min. The melting time for T = 1460 K is 20 min and for T = 1800 K is 25 min, respectively. The raw materials mass was about 500 g (each sample). The chemical composition of the copper samples before and after EBM is determined with ARL 4460 OES Thermo Scientific spectrometer (Thermo Fisher Scientific, Waltham, MA, USA). The spectrometer is equipped with a Paschen-Runge vacuum polychromator working in argon atmosphere. Oxygen analyses of the samples were performed with ELTRA OH-900 oxygen/hydrogen determinator (Eltra GMBH, Haan, Germany).

A 4% solution of nitric acid in ethyl alcohol was used to etch and reveal the microstructure of the obtained metal specimens. The etching time was 30 s.

A light microscope Leica DM2500 (Leica Microsystems GmbH, Wetzlar, Germany) with a digital camera Leica EC3 (Leica Microsystems GmbH, Germany) was used for the metallographic study of the macro or micro-structure of polished and etched surfaces of copper samples. The image processing was performed using the Leica LAS software (Leica Microsystems GmbH, Germany).

3. Results

3.1. Thermodynamic Analysis of Possible Chemical Interactions during Electron Beam Melting and Refining (EBMR)

The thermodynamic analysis of the possible chemical interactions occurring during the refining of copper from impurities such as Se, Te, Bi, As, Sb, Pb, Sn, Ni, Zn, Ag under EBM conditions is performed on the basis of the Gibbs free energy (ΔF) and the physical state of the impurities. The analysis was carried out using the professional thermochemical calculation programme HSC Chemistry ver.7.1, module "Reaction Equation" [21], taking into account the physical state of copper and the metal impurities during EBM.

Since there is a constant pressure in the vacuum chamber during EBM, the main parameters that affect the refining process are the temperature of the metal and its physical state [9]. Another parameter that impacts the removal of impurities from the main metal is the mass transport of molten or solid metal particles to the reaction surface [22].

Figure 2 shows the melting and boiling temperatures of studied metals and compounds. The temperature range 1460–1800 K of e-beam melting process is marked in Figure 2 by dashed lines.

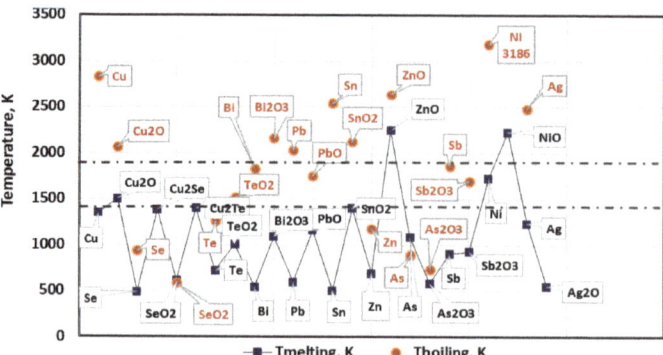

Figure 2. Melting and boiling temperatures of studied metals and compounds.

It can be seen that under vacuum conditions and the studied temperature range, the metal impurities present in Cu such as Pb, Bi, Sn, Sb, Ag and their oxides will be in a liquid state. Impurities such as Zn, As and its oxide have a boiling point significantly lower than the melting point of copper and they will be in the gaseous state. Figure 2 shows that the boiling points of Cu_2O, Pb and Bi_2O_3 slightly exceed the operating temperature of 1800 K, which allows us to assume that they will also be in gaseous state under vacuum conditions.

Phase diagrams Cu-Se and Cu-Te show that both impurities are present in copper in the form of intermetallic compounds: copper selenides and tellurides [23,24]. At low selenium and tellurium content they are in the form of Cu_2Se and Cu_2Te. These compounds will be present in liquid state as they have a low melting point.

Ni and Ag remain in liquid state. Nickel has complete mutual solubility in copper and at 1358 K (the melting temperature of copper) it completely passes into liquid phase [25,26]. The melting temperature of ZnO, unlike Ni, is very high and under e-beam melting conditions ZnO will be present in solid state.

Therefore, under EBM conditions, the liquid metal is a complex system of Cu, Cu_2O, metal impurities and their oxides, and they are in liquid, solid or gaseous state depending on the thermodynamic conditions of refining and the type of impurity.

Under e-beam melting conditions, the refining processes take place mainly on the reaction surfaces of the liquid metal (its interface with the vacuum, Figure 1) in three reaction zones [9]. Depending on the thermodynamic conditions of the EBM and the type of the individual impurities, the refining process can take place by: (i) degassing (removal of components, with a higher partial pressure than the partial pressure of the base metal), (ii) distillation (evaporation of the more volatile compounds from the metal components). Effective refining requires the implementation of the following inequalities concerning vapor pressures (p) of copper and the metallic impurities (R_i): $(p_{RiO}) > (p_{Ri}) > (p_{Cu_2O}) > (p_{Cu})$.

Thermodynamic evaluation of the possible chemical interactions in the system $Cu(l)$-$Cu_2O(l)$-$R_i(l,g)$ is made on the basis of the following equations:

$$2Cu(l) + [O] = Cu_2O(l) + \Delta F_{T,Cu/Cu_2O}, \quad (1)$$

$$R_i(l,g) + [O] = R_iO(l,g) + \Delta F_{T,Ri/RiO}, \quad (2)$$

$$Cu_2O(s,l) + R_i(l,g) = R_iO(l,g) + 2Cu(l) + \Delta F_T, \quad (3)$$

where $\Delta F_{T,Cu/Cu_2O}$, $\Delta F_{T,Ri/RiO}$ and ΔF_T are the Gibbs free energies of the respective processes. The indices (s), (l) and (g) mean that the substance is in a solid, liquid or gaseous state, respectively. The calculations were performed under melting conditions: at temperatures of 1460 K, 1500 K, 1600 K, 1700 K and 1800 K and operating pressure in the vacuum chamber of 10^{-3} Pa. These parameters correspond to the actual conditions of melting and refining of copper technogenic material in the EBM plant.

Figure 3 shows the temperature dependences of the free energies of oxidation of copper ($\Delta F_{T,Cu/Cu_2O}$) and metal-impurities ($\Delta F_{T,Ri/RiO}$). The analysis of the obtained dependences shows that the probability of oxidation of ZnO(g) to ZnO(s) is the highest and as the temperature increases, the affinity of Zn(g) to oxygen decreases significantly. In the whole temperature range, the values of ΔF of formation of $SnO_2(l)$, $Sb_2O_3(l)$, $As_2O_3(l)$ and PbO(l) are higher than that of the oxidation of Cu(l) to $Cu_2O(l)$. The thermodynamic probability of oxidation of bismuth to $Bi_2O_3(l)$ is almost the same as that of copper to Cu_2O, while the probability of oxidation of copper telluride to $TeO_2(l)$ and copper selenide to $SeO_2(l)$ is significantly lower than that of copper.

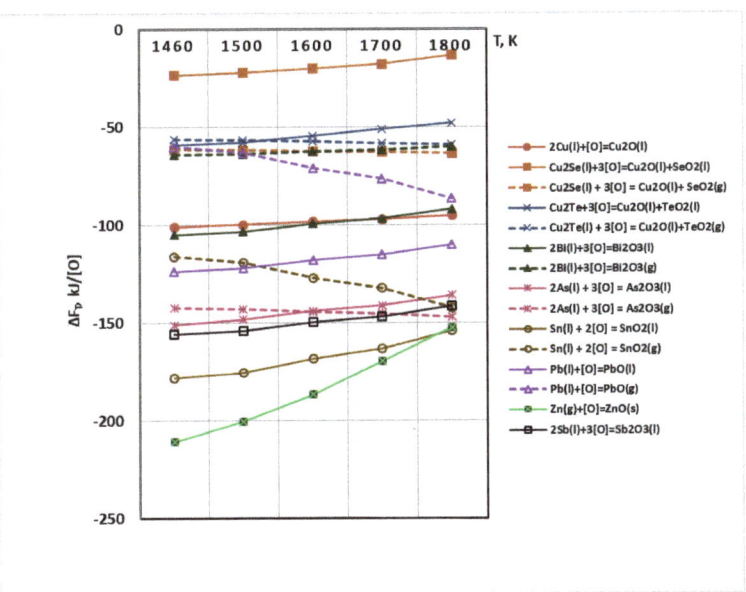

Figure 3. Influence of the temperature on ΔF_T of oxidation reactions of Cu(l) and R_i(l,g) under vacuum conditions.

It is observed that as the temperature increases, the values of the Gibbs energy of the oxidation of impurities: Sn(l) → SnO_2(g) and Pb(l) → PbO(g) increase significantly but they are significantly lower than those of oxidation to liquid oxides.

Energy of oxidation of Bi(l) → Bi_2O_3(g), As → As_2O_3(g), Cu_2Se → SeO_2(g) and Cu_2Te → TeO_2(g) is almost independent of temperature.

Out of the impurities present in copper, only Cu_2Se shows a higher thermodynamic probability of oxidizing to SeO_2(g) rather than to SeO_2(l). This trend is also observed in Cu_2Te when increasing the temperature above 1500 K.

The possibility of chemical interactions between Cu_2O(s,l) and metal impurities is described by Equation (3). The calculated values of ΔF_T are shown in Figure 4. The calculations at T = 1460 K were performed under the condition that the copper oxide is in a solid state (Cu_2O(s)) as the melting temperature of the copper oxide is higher (1508 K).

It is observed that in the studied temperature range all impurities will interact with Cu_2O regardless of its phase state. The calculated values of ΔF_T are high, which indicates a high thermodynamic probability of the process. The chemical interaction of copper oxide with gaseous zinc is most likely to occur in this case as well. Zinc is oxidized to a stable ZnO compound which floats to the reaction surface and Zn will be separated by degassing [20].

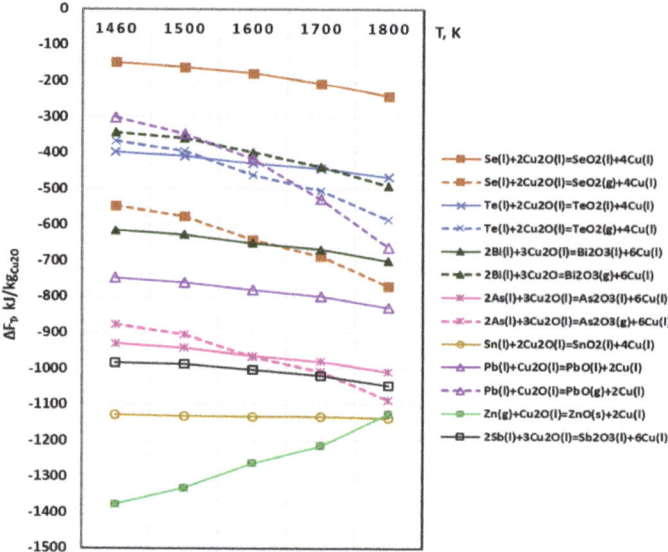

Figure 4. Free energy for interaction between Cu_2O and metals impurities in vacuum.

The oxidation reactions of Ni and Ag impurities are not shown in Figures 3 and 4, as the calculated values of ΔF_T are positive within the investigated temperature range. According to the thermodynamic laws, the course of a given reaction is possible only when the calculated value of the free energy is negative ($\Delta F_T < 0$) [27].

The analysis of the reactions with formation of gaseous phase shows that at temperatures above 1600 K, Se, Te and As will be removed mainly in gaseous state, while Sn, Sb, Pb, Bi will be oxidized mainly to oxides–$SnO_2(l)$, $Sb_2O_3(l)$, $PbO(l)$, $Bi_2O_3(l)$.

Following the performed thermodynamic analysis, it can be concluded that in the studied temperature range impurities such as Se, Te and As will be oxidized to gaseous oxides, while Sn, Sb, Pb, Bi will be oxidized mainly to liquid oxides.

3.2. Refining Efficiency and Microstructures of Obtained Copper

The influence of the temperature (beam power) and the duration of the retention time (τ), during which the melting metal is in liquid state, on the degree of removal of the impurities present in copper technogenic material is evaluated as well. Data about chemical analysis of the impurities concentration of the starting copper material (before EBMR) and of the specimens after e-beam refining of Cu, material losses (estimated using the weight of the initial material and the obtained ingots) and structure of the melted samples are obtained and analyzed under each of the technological regimes studied.

Data for the material losses (W_{loss}) which are mainly due to evaporation and also to splashes is presented in Table 1. The results show that the increase of the beam power (temperature) and also the increase of the residence time (τ) lead to an increase of the weight losses W_{loss} (Table 1). The minimum weight loss is 1.63% at T = 1700 K for τ = 20 min.

Data about the changes in the chemical composition of the samples after melting under different EBM technological parameters (regimes) is also presented in Table 1.

The influence of the temperature on the degree of impurity removal (α) is presented in Figure 5. The values of the degree of impurity removal are calculated from:

$$\alpha(i) = \frac{C_{Ri(initial)} - C_{Ri(final)}}{C_{Ri(initial)}} \cdot 100\% \qquad (4)$$

where $C_{Ri(initial)}$ and $C_{Ri(final)}$ are the initial and final impurity concentrations, respectively.

Table 1. Concentration of impurities at electron beam processing of copper technogenic material.

Probe	Cu-00	Cu-01	Cu-02	Cu-03	Cu-04	Cu-05	Cu-06	Cu-07
Type	Concentration before EBMR ppm	T = 1460 K t = 20 min	T = 1500 K t = 15 min	t = 35 min	T = 1600 K t = 20 min	T = 1700 K t = 20 min	t = 45 min	T = 1800 K t = 25 min
Se, ppm	382	367	331	309	143	105	107	103
Te, ppm	68	66	55	56	12	10	10	11
Bi, ppm	150	122	74	73	<10	<1	<1	<1
Pb, ppm	210	115	87	82	<3	<3	<1	<1
Sn, ppm	34	24	16	18	<2	<2	<1	<1
Sb, ppm	95	60	45	32	15	<9	<4	<3
As, ppm	700	577	416	403	106	50	50	<3
Ni, ppm	1000	870	740	750	750	660	620	580
Zn, ppm	23	11	8	6	<1	<1	<1	<1
Ag, ppm	140	135	133	131	130	131	131	132
O, ppm	2251	897	250	50	15	10	10	10
S, ppm	28	14	7	5	<1	<1	<1	<1
Cu,%	99.454	99.675	99.784	99.809	99.892	99.902	99.907	99.916
[1] ε_{tot},%		40.5	60.4	65.0	78.3	82.1	83.0	84.6
W_{loss},%		0.42	0.86	2.07	1.63	2.93	3.56	3.06

[1] ε_{tot}-total efficiency of copper refining.

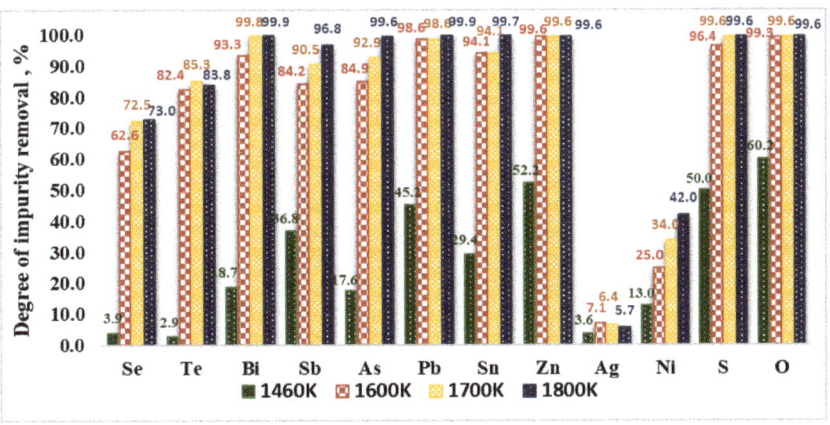

Figure 5. Influence of the temperature on the removal of impurities from copper technogenic material at EBMR.

It can be observed that increasing the temperature from 1460 K to 1600 K leads to the intensive removal of impurities such as Bi, Pb, Sn, Zn and As. The degree of removal of these impurities with the exception of As is more than 93%. In this case, the degree of oxygen removal is 99.3%. This means that during EBM the removal of impurities takes place mainly as a result of the chemical interaction with Cu_2O. The lower removal values of As (84.9%) and Sb (84.2%) can be explained by the formation of complex compounds between these impurities and Ni (such as strong chalcophyllite $3Cu_2O \cdot 4NiO \cdot Sb_2O_5$ and antimony arsenate $Sb_2O_3 \cdot As_2O_5$) during EBMR of copper.

The removal rates (degree of refining) of Se and Te at 1600 K are 62.6% and 82.4%, respectively which can be explained by the fact that these impurities form intermetallic compounds with copper, which are more difficult to oxidize. This fact is consistent with the significantly lower values of the Gibbs energy of the oxidation reactions of these impurities (Figures 3 and 4).

The removal values of oxygen and sulphur at 1600 K are 99.33% and 96.43%, respectively and they increase to 99.56% and 99.64% at 1800 K. Silver and almost all of the nickel remain in copper. Nickel losses can be explained by the total copper losses.

The influence of the refining time on the degree of impurity removal (α) is evaluated for 1500 K and 1700 K and the calculated values are presented in Figure 6. The analysis of the results shows that at both temperatures, extending the refining time to more than 15–20 min does not significantly affect the degree of removal of impurities from the technogenic copper material. It can be observed that at a temperature of 1700 K and a duration of 20 min the degree of removal of Bi, Pb, Zn is over 98% and that of As, Sb, Sn-about 91–94%. The removal rates of Se (72.5%) and Te (85.3%) are lower. A significant part of Ni and Ag remains in copper.

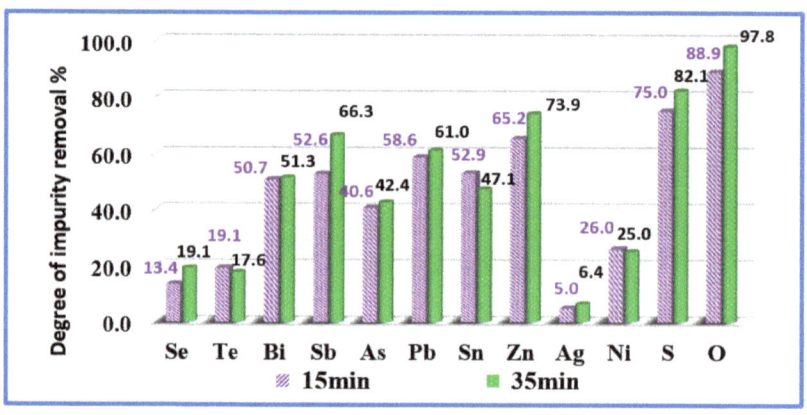

(a)

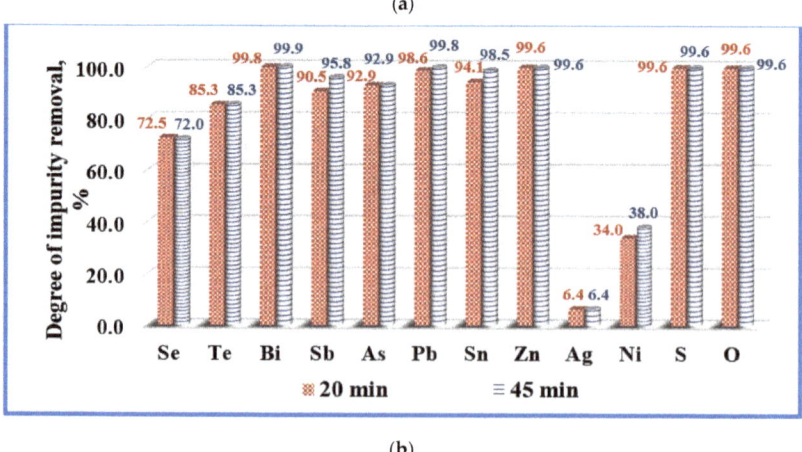

(b)

Figure 6. Influence of the melting time on the removal of impurities from copper technogenic material at temperatures: (a) T = 1500 K (samples Cu-02 and Cu-03); and (b) T = 1700 K (samples Cu-05 and Cu-06).

The highest removal efficiency of oxygen is 99.6% (the minimal oxygen content of 10 ppm) and is obtained at T = 1700 K and T = 1800 K (Figures 5 and 6, Table 1). The optimum of the oxygen refining is connected to higher superheating of the molten metal and better reduction of the oxygen content independently from the retention time in the molten state of the refining copper.

At a temperature of 1500 K extending the refining time increases the rate of sulphur removal from 75% to 82.1%. At higher temperatures, the removal rate is 99.6% regardless of the retention time. The same trend is observed with oxygen.

Figure 7 shows microstructures of the Cu-02 sample manufactured at a temperature of 1500 K for 15 min retention time. The presented structures are from the upper surface of the ingot (top of the ingot)-Cu-02(t) and from the surface along the depth of the ingot (transverse section)-Cu-02(s).

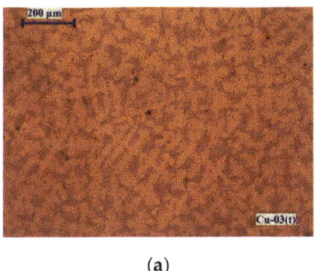
(a)

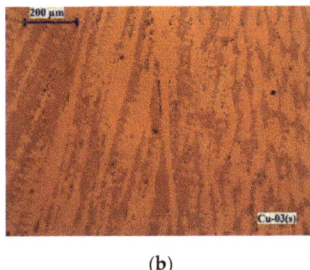
(b)

Figure 7. Macrostructures of sample Cu-02 obtained at T = 1500 K for τ = 15 min: (**a**) from the top of the ingot; (**b**) from the transverse section.

The microstructure observed on the upper surface of the sample (Figure 7a) is dendritic and shows the formation of eutectic melts of type $E(Cu-Cu_2O)$, $E(Cu-Cu_2S)$, the presence of loose eutectic melts of insoluble in copper impurities (Pb, Bi) and crystallization of intermetallic phases of Se and Te. The dark stripes in the micrograph of the transverse section of the sample (Figure 7b) show the direction of their crystallization in the volume of copper.

The effects of the e-beam power and melting time on the microstructures of the copper ingots are presented in Figure 8.

By increasing the beam power (temperature) for 20 min refining time (Figure 8a–d), formation of well-formed globulitic crystals, which are characteristic of nickel-containing copper, is observed. The presence of copper-soluble impurity Ag does not affect the structure. The micrographs in Figure 8d,e show that the extending of the refining time at a given temperature does not affect the microstructure. In this case, the impurity removal rates have similar values.

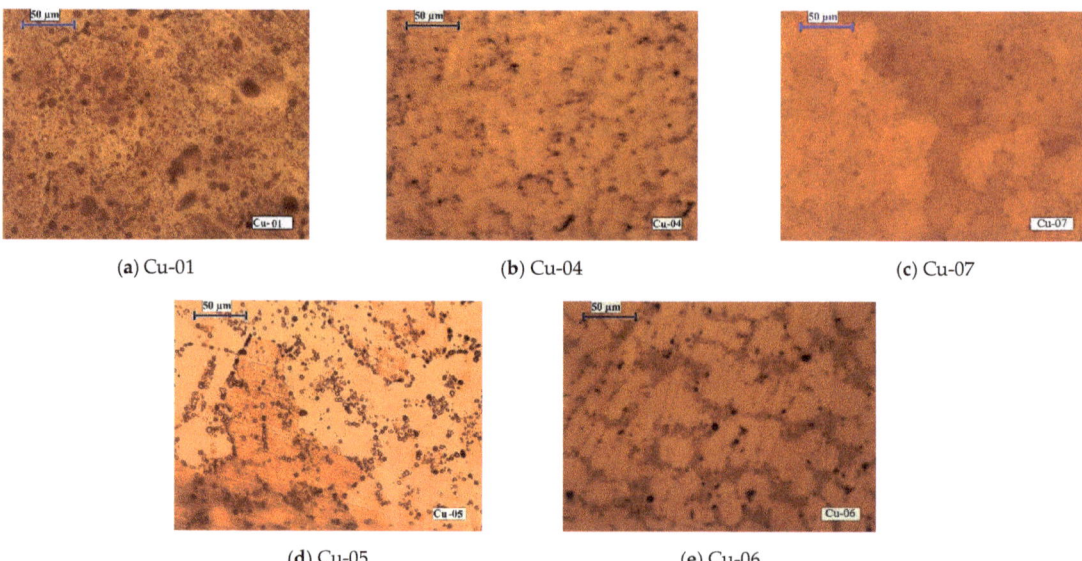

Figure 8. Macrostructures of copper samples manufactured at different EBMR conditions: (**a**) P_b = 6 kW, τ = 20 min; (**b**) P_b = 10 kW, τ = 20 min; (**c**) P_b = 19.5 kW, τ = 20 min; (**d**) P_b = 13 kW, τ = 20 min; and (**e**) P_b = 13 kW, τ = 45 min.

4. Conclusions

The paper investigates the possibility of removing impurities from technogenic copper material (99.45%) using EBM. On the basis of the thermodynamic analysis of the possible chemical interactions occurring in the Cu-Cu_2O-R_x system in the studied temperature range 1460–1800 K and the conducted experimental studies, the influence of the kinetic parameters-temperature (beam power) and melting time on the degree of removal of non-metallic (O, S) and metallic (Se, Te, Pb, Bi, Sn, As, Sb, Zn, Ni, Ag) impurities, the refining efficiency and the structure of the resulting copper was evaluated. The conclusions can be summarized as follows:

- The results obtained show that the electron beam melting method can be successfully applied for refining copper technogenic material with a high content of impurities, which in the conditions of EBM are in a gaseous state (such as Bi, Pb, Zn, As, Sb, Sn) and reach nearly 100% removal degree and ~97% for Sb.
- Oxygen and sulphur also reach a high degree of removal ($\geq$99%). Under the studied conditions, the maximum degree of refining of Se and Te is 73% and 85.3%, respectively, which is due to the fact that Se and Te form intermetallic compounds with copper, which are more difficult to oxidize. This corresponds to significantly lower values of the Gibbs energy of the oxidation reactions of these impurities.
- Silver and most of the nickel remain in copper. Under vacuum conditions and at the temperature range studied, silver does not oxidize or evaporate. The low degree of refining of nickel (34–42%) from copper can be explained by the good solubility of this impurity in copper. In addition, nickel is also not oxidized in the studied temperature range. It was found that raising the temperature above 1700 K, as well as extending the melting time over 20 min hardly change the purity and structure of the resulting refined metal.
- At temperatures in the range 1600–1800 K, the achieved refining efficiency is 78–85% and the purity of copper after EBM is 99.9%. The highest total refining efficiency of 84.6% is seen at a beam power of 19.5 kW for 20 min melting time and the best purification of copper technogenic material (99.92%) is achieved.

Author Contributions: K.V. and V.S. contributed to the design of the study and interpretation of data; K.V. and V.V. conceived and designed the experiments; V.V. performed the experiments; K.V., V.S. and M.K. analyzed the data; K.V. and V.S. wrote the manuscript. All authors have read and agreed to the published version of the manuscript.

Funding: The work was supported by the Bulgarian National Science Fund under contract DN17/9.

Institutional Review Board Statement: Not applicable.

Informed Consent Statement: Not applicable.

Acknowledgments: The authors are grateful to M. Naplatanova and R. Nikolov for technical assistance in processing the samples.

Conflicts of Interest: The authors declare no conflict of interest.

References

1. International Copper Study Group. *The World Copper Factbook*; 2020. Available online: http://www.icsg.org/index.php/component/jdownloads/finish/170/3046 (accessed on 23 January 2022).
2. European Copper Institute Copper: A Circular Material for a Resource Efficient Europe. Available online: https://copperalliance.eu/benefits-of-copper/recycling/ (accessed on 3 December 2021).
3. Biswas, A.K.; Davenport, W.G. *Extractive Metallurgy of Copper*, 3rd ed.; Elsevier: Amsterdam, The Netherlands, 2013; p. 518.
4. Hanusch, K.; Bussmann, H. Behavior and removal of associated metals in the secondary metallurgy of copper. In Proceedings of the 3rd International Symposium on Recycling of Materials and Engineered Materials, Point Clear, AL, USA, 12–16 November 1995; pp. 171–188.
5. Antrekowitsch, H.; Wenzl, C.; Filzwieser, I.; Offenthaler, D. *Pyrometallurgical Refining of Copper in an Anode Furnace*; TMS, Ed.; The Minerals, Metals & Materials Society: Warrendale, PA, USA, 2005; pp. 1–6.
6. Asghari, H.; Safarzadeh, M.S.; Asghari, G.; Moradkham, D. The effect of impurities on the extraction of copper from sulfate medium using LIX®984N in kerosene. *Russ. J. Non-Ferr. Met.* **2009**, *50*, 89–96. [CrossRef]
7. Navarro, P.; Vargas, C.; Castillo, J.; Sepulveda, R. Experimental Study of Phase Entrainment in Copper Solvent Extraction. *DYNA* **2020**, *87*, 85–90. [CrossRef]
8. Mackey, P.J.; Wraith, J. Development of copper quality: An historical perspective. *Miner. Process. Extr. Metall* **2004**, *113*, 25–37. [CrossRef]
9. Mladenov, G.; Koleva, E.; Vutova, K.; Vassileva, V. *Practical Aspects and Application of Electron Beam Irradiation*; Memtanu, M., Brasoveanu, M., Eds.; Transworld Research Network: Trivandrum, India, 2011; pp. 43–93.
10. Kalugin, A. *Electron Beam Melting of Metals*; Metallurgy Publishing House: Moscow, Russia, 1980. (In Russian)
11. Mitchell, A.; Wang, T. Electron beam melting technology review. In Proceedings of the Conference Electron Beam Melting and Refining, Reno, NV, USA, 29–31 October 2000; pp. 2–13.
12. Vassileva, V.; Mladenov, G.; Vutova, K.; Nikolov, T.; Georgieva, E. Oxygen removal during electron beam drip melting and refining. *Vacuum* **2005**, *77*, 429–436. [CrossRef]
13. Vutova, K.; Vassileva, V.; Koleva, E.; Munirathnam, N.; Amalnerkar, D.; Tanaka, T. Investigation of Tantalum Recycling by Electron Beam Melting. *Metals* **2016**, *6*, 287. [CrossRef]
14. Yue, H.; Peng, H.; Li, R.; Gao, R.; Wang, X.; Chen, Y. High-temperature microstructure stability and fracture toughness of TiAl alloy prepared via electron beam smelting and selective electron beam melting. *Intermetallics* **2021**, *136*, 107259. [CrossRef]
15. You, X.; Tan, Y.; Cui, H.; Zhang, H.; Zhuang, X.; Zhao, L.; Niu, S.; Li, Y.; Li, P. Microstructure evolution of an Inconel 718 alloy prepared by electron beam smelting. *Mater. Charact.* **2021**, *173*, 110925. [CrossRef]
16. Vutova, K.; Vassileva, V.; Stefanova, V.; Amalnerkar, D.; Tanaka, T. Effect of electron beam method on processing of titanium technogenic material. *Metals* **2019**, *9*, 683. [CrossRef]
17. Yazawa, A.; Nakazawa, S.; Takeda, Y. Distribution behavior of various elements in copper smelting Systems. In Proceedings of the International Sulfide Smelting Symposium: Advances in Sulfide Smelting, San Francisco, CA, USA, 6–9 November 1983; Volume 1, pp. 99–117.
18. Kamberovic, J.; Ranitovic, M.; Korac, M.; Jovanovic, N.; Tomovic, B. Pyro-Refining of Mechanically Treated Waste Printed Circuit Boards in DC Arc-Furnace. *J. Sustain. Metall.* **2018**, *4*, 251–259. [CrossRef]
19. Klementtinen, L.; Avarmaa, K.; O'Brien, H.; Taskinen, P. Behavior of Tin and Antimony in Secondary Copper Smelting Process. *Minerals* **2019**, *9*, 39. [CrossRef]
20. Vassileva, V.; Vutova, K. Influence of process parameters on quality of copper in electron-beam melting. *J. Phys. Conf. Ser.* **2020**, *1492*, 012014. [CrossRef]
21. Roine, A.; Kotiranta, T.; Esrola, K.; Lamberg, P. *HSC Chemistry v. 7.1*; Metso Outotec: Helsinki, Finland, 2011.
22. Vassileva, V.; Vutova, K.; Mladenov, G. Analysis of the Thermodynamics of Refining during Electron Beam Melting of Refractory Metals. *Mater. Und Werkst.* **2006**, *37*, 613–618. [CrossRef]

23. Phase Diagram of the Cu-Se System. Available online: http://www.himikatus.ru/art/phase-diagr1/Cu-Se.php (accessed on 23 January 2022).
24. Phase Diagram of the Cu-Te System. Available online: http://www.himikatus.ru/art/phase-diagr1/Cu-Te.php (accessed on 23 January 2022).
25. Phase Diagram of the Cu-Ni System. Available online: http://www.himikatus.ru/art/phase-diagr1/Cu-Ni.php (accessed on 23 January 2022).
26. Kawecki, A.; Knych, T.; Sieja-Smaga, E.; Mamala, A.; Kwasniewski, P.; Kiesiewicz, G.; Smyrak, B.; Pacewicz, A. Fabrication, properties, and microstructures of high strength and high conductivity copper-silver wires. *Arch. Metall. Mater.* **2012**, *57*, 4. [CrossRef]
27. Linchevskiy, B.V. *Thermodynamics and Kinetics of the Interaction of Gases with Liquid Metals*; Metallurgy Publishing House: Moscow, Russia, 1986. (In Russian)

Cathodoluminescent Analysis of Sapphire Surface Etching Processes in a Medium-Energy Electron Beam

Arsen Muslimov * and Vladimir Kanevsky

Shubnikov Institute of Crystallography, Federal Scientific Research Center Crystallography and Photonics, Russian Academy of Sciences, 119333 Moscow, Russia; kanevsky@crys.ras.ru
* Correspondence: amuslimov@mail.ru

Abstract: Sapphire crystals are widely used in optics and optoelectronics. In this regard, it is important to study the stability of crystals under external influence and the possibility of modifying their surfaces by external influence. This work presents the results of studying the processes of the action of an electron beam with an average energy of 70 keV or less under vacuum conditions on the surfaces of sapphire substrates of various orientations. The effect of etching a sapphire surface by an electron beam in vacuum at room temperature was discovered. The highest etching rate was observed for A-plane sapphire (the average pit etching rate was 10^{-6} μm^3/s). It was shown that the rate of etching of a sapphire surface increased many times over when gold is deposited. An in situ method for studying the process of etching a sapphire surface using cathodoluminescence analysis was considered. Possible mechanisms of sapphire etching by a beam of bombarding electrons were considered. The results obtained could be important in solving the problem of the stability of sapphire windows used in various conditions, including outer space. In addition, the proposed method of metal-stimulated etching of a sapphire surface can be widely used in patterned sapphire substrate (PSS) technology and further forming low-dislocation light-emitting structures on them.

Keywords: patterned sapphire substrate; electron etching; gold; cathodoluminescent analysis; anisotropy; light-emitting diodes; windows

1. Introduction

Sapphire crystals (corundum, α–Al$_2$O$_3$), because of the peculiarities of their chemical composition and crystal structure, are able to withstand a range of intense external influences, including high temperature, pressure, and mechanical stress. Sapphire belongs to the trigonal (rhombohedral) crystal system, as a result of which it has a pronounced anisotropy of physical and chemical properties. Sapphire crystals are used mainly in optics and optoelectronics, where the anisotropy of properties is most in demand. For example, for high-strength optical glasses, it is preferable to use C(0001)-plane sapphire with zero birefringence [1]. Furthermore, the C-plane of sapphire has been used for a long time as a substrate for the formation of light-emitting devices [2–4]. Thus, using the C-plane sapphire profiling technique, it was possible to increase the light yield of devices by up to 20% [5,6]. However, it was revealed that there is a fundamental problem associated with the use of C-plane sapphire in light-emitting devices, which prevents a further increase in efficiency: the quantum-confined Stark effect [7,8]. The presence of near-surface electric fields in polar structures leads to a loss of efficiency and a long-wavelength shift of the radiation maximum. It is possible to reduce or completely suppress the effect of near-surface electric fields by using semipolar or nonpolar oriented nitride layers. In particular, such structures may be formed on M-plane (10$\bar{1}$0), R-plane (1$\bar{1}$02), and A-plane (11$\bar{2}$0) sapphire [9–12].

There are several important application requirements for sapphire crystals in general. On the one hand, they are associated with the use of devices operated in a wide variety of environments, including space flight systems, as protective optical glasses. It should be borne in mind that apparatuses are susceptible to various types of ionizing radiation and flows of charged particles of various energies in vacuum conditions. On the other hand, optimization of the pregrowth preparation of the substrate surface (polishing, microstructuring, profiling) is also important in the formation of light-emitting devices. Note that these requirements are related: methods of external action that lead to decomposition or etching of the surface of sapphire plates can be considered as methods of microstructuring and profiling their surface.

A number of works [13–15] have been devoted to the study of the processes of influence of ionic flows leading to the transformation of the surfaces of sapphire substrates. The effects of various types of radiation on sapphire optical fibers with a total radiation dose of up to 3.39 mGy has been more widely studied [16]. The effect of electrons is also used in electron lithography for resist processing [17] and in high-energy electron accelerators [18]. In this paper, we present the results of studying the processes of the action of an electron beam with an average energy of 70 keV or less under vacuum conditions on the surfaces of sapphire substrates of various orientations.

Since the processes of surface transformation in the case of external action usually cover the near-surface region of crystals up to tens of microns, contact research methods are not applicable. The only possibility is the use of noncontact methods based on the analysis of electromagnetic radiation from the surface under investigation [19,20]. In particular, the authors of [19] demonstrated the possibility of in situ studies of the processes of transformation of sapphire surface using expensive synchrotron radiation sources. In this paper, we considered the possibility of in situ study of the process of etching the sapphire surface as a result of medium-energy electron bombardment using luminescence analysis methods. This possibility appeared to be due to features of the cathodoluminescence (CL) spectra of sapphire crystals. Anionic defects in sapphire crystals are identified as F- and F^+-centers, which are oxygen vacancies that have captured two electrons and one electron, respectively, and their complexes. In the ultraviolet (UV) region of the sapphire CL spectrum, bands are observed in the regions of 330–340 nm (F^+-center) and 410–415 nm (F-center) [21]. Since the F- and F^+-centers are exclusively intrinsic defects of the crystal structure, their concentration is directly dependent on the structural-phase composition of the crystal. On the contrary, in the long-wavelength part of the sapphire CL emission spectrum, bands associated with Cr and Ti atoms with impurities are traditionally observed, which are less dependent on the structural-phase composition and can appear even in the molten state of Al_2O_3. In particular, a narrow line at 694 nm and a broad band peaking at 780 nm are associated with ^2E–$^4A_2$ and ^2E–$^2T_2$ transitions in Cr^{3+} and Ti^{3+} [22], respectively. Interestingly, heating or increasing the fluence increases the luminescence intensity in the UV region of the sapphire emission spectrum, while the dependence on electron energy has a maximum [23], after which a decrease is observed. The broadening of the chromium line in the long-wavelength region corresponds to an increase in the temperature of shock-shifted luminescent chromium ions above the stationary temperature of the crystal [24]. A similar broadening is observed for a broad band associated with Ti^{3+} [25]. In general, it is possible to study the processes of sapphire etching using the basics of quantitative luminescent analysis, which assume a linear relationship between the luminescence intensity and the concentration of luminescence centers at low concentrations.

Thus, the independent behavior of color centers of different nature during external action makes it possible to consider cathodoluminescence analysis as a very promising in situ method for studying the processes of rearrangement and etching of the surface of sapphire crystals.

2. Materials and Methods

The samples were C-, A-, M-, and R-plane sapphire with one-sided chemical-mechanical polishing. It should be noted that the content of chromium and titanium impurities with a concentration of ~10 ppm found in the sapphire crystal was due to their presence in the initial charge. Two types of wafers were used: (a) initial and (b) after annealing at a temperature of 1400 °C for an hour under atmospheric conditions to recrystallize the surface and reduce the concentration of surface-layer defects introduced during growth, machining, and polishing. Next, a layer of gold about 100 nm thick was formed on an A-plane of sapphire by thermal vacuum deposition (VN-2000 configuration). The sample was then annealed in air Naber tube furnace (Nabertherm, Lilienthal, Germany) for 2 h at 800 °C to form a discrete gold structure.

The study by the method of excitation of CL in the samples was carried out with an electron beam of an EG-75 electron diffraction recorder, the electron energy of which was 40 keV, 50 keV, 60 keV, and 70 keV; the spot diameter of which was 0.5 and 3 mm; the electron flux density of which was 10^{21} cm$^{-2} \cdot$s^{-1}; and the electron beam current of which was 80 µA. This was undertaken at room temperature (special heating of the samples was not performed). The CL spectra of all samples were studied 10 s after the start of irradiation. The vacuum was maintained in the 10^{-4} Pa system. We used AvaSpec-ULS2048x64-USB2 spectrophotometric complex (Avantes). The angle of incidence of the electron beam on the plane of the substrate was 45°, and the angle between the axis of the fiber-optic adapter and the direction of propagation of the incident electron beam was 90°. The time dependence of the CR spectra was studied with the following exposure parameters: numbers of spectra, 21 and 46; time interval between spectra, 8 s.

Microscopic studies of the surface of the samples were carried out on an SEM Leo-1450 scanning electron microscope (Carl Zeiss AG, Oberkochen, Germany) and an Ntegra Aura atomic force microscope (NT-MDT, Zelenograd, Russia) in the modes of semicontact topography and phase contrast, respectively.

3. Results and Discussion

3.1. Orientation Dependence of the CL Spectra of Sapphire Samples

Figure 1a shows the CL spectra (electron energy 50 keV, spot diameter 0.5 mm) of sapphire samples of various orientations after heat treatment at 1400 °C under atmospheric conditions. The spectra showed an intense main F$^+$(340 nm)-band, but the F(415 nm)-band was suppressed. In this case, an effect similar to that previously found in [21] was observed. When irradiated with electrons, the F-centers were suppressed because of the formation of new F$^+$-centers according to:

$$F + \text{exposure} \rightarrow (F^+)^* + e^-_{\text{trap}} \rightarrow F^+ + h\nu\,(330\text{ nm}) + e^-_{\text{trap}} \quad (1)$$

where $(F^+)^*$ is an excited F$^+$-center.

According to Figure 1a, the absolute intensities I of the luminescence of the CL F$^+$-bands were related as follows: $I_{\text{A-plane}} > I_{\text{M-plane}} > I_{\text{R-plane}} > I_{\text{C-plane}}$. This corresponds to the ratio of the rates of ion etching of sapphire wafers of different orientations [14] at an accelerating voltage of 30 kV. In both cases, the etching rate was maximum for A-plane sapphire and minimum for C-plane sapphire. Additionally (Figure 1b), the luminescence intensity of the F$^+$-band of the CL spectrum was studied for the initial and heat-treated A-plane and C-plane samples (for which the maximum and minimum CL intensities were observed, according to the data in Figure 1a) at a higher electron energy (60 keV). With an increase in energy, electrons were able to penetrate into deeper layers and generate oxygen vacancies in them. Figure 1b demonstrates that the general trend continued: $I_A > I_C$.

Passing from the luminescence spectra to the processes of electron etching of the sapphire surface, it should be kept in mind that the sapphire substrates were preliminarily annealed to minimize the concentration of oxygen vacancies, and we can assume that the CL intensity in the UV region of the spectrum was proportional to the concentration of oxygen vacancies. In addition, the results showed that an increase in the accelerating voltage led to an increase in the intensity of the emission of F$^+$-centers (oxygen vacancies that had captured one electron). In this case, the glow of F-centers (oxygen vacancies that had captured two electrons) was suppressed. Considering that the number of electrons emitted by the cathode every second did not change, the increase in the intensity of the glow of the F$^+$-centers could have been associated only with the generation of new oxygen vacancies from deep layers.

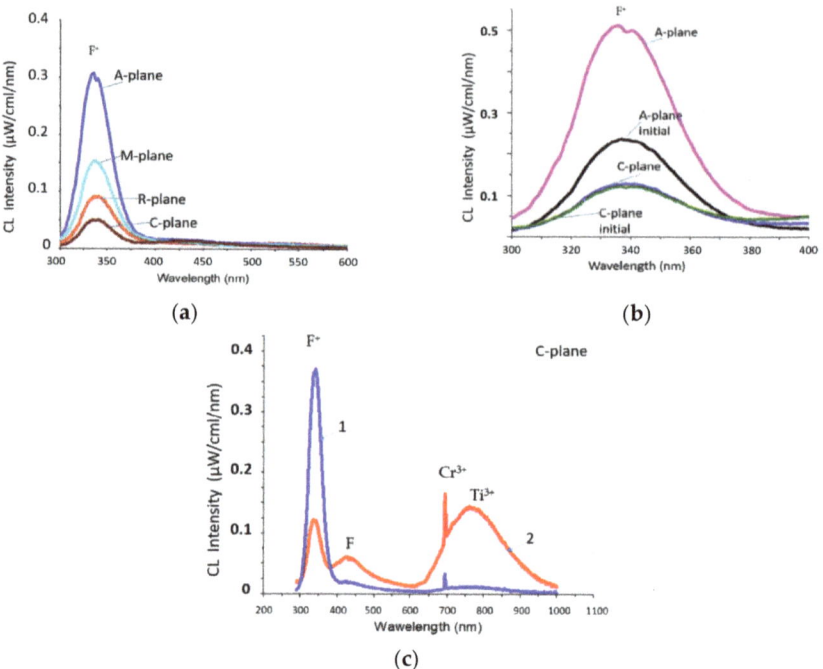

Figure 1. CL spectra of heat-treated at 1400 °C and initial sapphire samples at electron energies of 50 keV (**a**) and 60 keV (**b**). Dependence of the CL spectrum (electron energy 70 keV) of C-plane sapphire on the spot diameter (**c**): 1–0.5 mm; 2–3 mm.

The correspondence of the etching processes of a sapphire surface by ion and electron flows, which we found in our experiments, indicated a universal mechanism of defect formation caused by the displacement of atoms during elastic collisions with bombarding particles. In the case of impact mixing, an important factor is the interaction of atoms in the atomic layer. According to accepted calculations [14], the atomic arrangement density is maximum in the C-plane and minimum in the A-plane: 25 and 7.63 at/nm^2, respectively. The potential energies per atom (the energy of removal of atoms) are 291.23 and 908.63 eV for the A-plane and C-plane, respectively. This explains the maximum rate of ion and electron etching of the A-plane of sapphire. On the contrary, the interplanar distances for the C-plane and A-plane are 2.165 and 1.190 Å, respectively. The binding energy of neighboring atomic layers along any direction in the crystal is inversely proportional to the corresponding interplanar distance. The rate of chemical–mechanical etching depends on the interplanar distance, and since this distance is maximum along C[0001], the etching rate is maximum. The hardness is minimal for C-plane sapphire.

Analysis of the CL spectra of the original and heat-treated samples (Figure 1b) showed a strong difference for A-plane sapphire but similarity for C-plane sapphire. According to [26], after chemical–mechanical polishing, the A-plane and R-plane of sapphire have the highest roughness, and the C-plane of sapphire, the lowest. The low intensity of the F$^+$-band for the original A-plane sapphire is associated with a high surface roughness. A significant part of the bombarding electrons is scattered by surface inhomogeneities without the formation of oxygen vacancies. After heat treatment, the surface becomes atomically smooth, the scattering of electrons by inhomogeneities decreases, and the rate of generation of oxygen vacancies increases. On heat-treated A-plane sapphire (Figure 1), there was a bifurcation of the F$^+$-peak, which was associated with the presence of luminescence components along and perpendicular to the main optical C-axis of the sapphire crystal. On the initial samples, the splitting of the F$^+$-peak was also not observed because of the strong scattering of electrons by surface defects. On the contrary, there was a significant similarity between the CL spectra of the original and heat-treated C-plane sapphire (Figure 1b). The basic (C-plane) sapphire is characterized by close-packed oxygen layers, and it is these layers that determine the features of the corundum-type crystal structure.

An interesting result was obtained in studying the dependence of the spectral features of CL on the diameter of the electron spot on sapphire samples. Figure 1c shows a typical picture of the evolution of the CL spectrum as the electron spot expanded from 0.5 mm to 3 mm. Bands in the UV region underwent radical changes: the F$^+$-band undergoes a reverse transition to the F-band according to the equation $F^+ + e^- \rightarrow F$. It is important to note in Figure 1c a weak Ti^{4+} peak (435 nm, curve-2) next to the F-peak on the right, which is detected when the spot expands to 3 mm. This peak, according to [27], is associated with F centers in sapphire: radiation stimulated capture of an electron by the F$^+$-center with Ti^{3+} to form the F-Ti^{4+} complex. In addition, the intensity of the bands in the long-wavelength region increases sharply (transitions in Cr^{3+} and Ti^{3+}). Moreover, the increase in the intensity of the Cr^{3+} and Ti^{3+} bands was proportional to the increase in the spot diameter. It can be assumed that upon defocusing (an increase in the diameter of the electron beam), saturation of the active Ti^{3+} and Cr^{3+} ions was not achieved, and the number of emitting centers was proportional to the cross-sectional area of the beam of electrons that excited luminescence.

To analyze the etching processes of the sapphire surface, the time dependence of the CL spectra was studied (Figure 2). The electron energy was assumed to be as low as 40 keV to reduce the electron beam instability error. The spot diameter was 0.5 mm. The intensity of the F$^+$-band, which was proportional to the concentration of oxygen vacancies, reached saturation level after the initial amplification. This was due to a set of processes occurring on the sapphire surface during irradiation, which were determined by the cross-sections of electron scattering, ionization, capture, and excitation, as well as the recombination and clustering of defects [23]. For the rest of the bands, a slight increase was observed over the entire interval of the study. The amplification of the bands could be associated with the process of partial etching of the sapphire surface and an increase in the number of active unsaturated luminescence centers. Thus, in the process of irradiation, an effect similar to the defocusing (diameter increase) of the electron spot was observed. The increase in the surface area was small, and therefore, the enhancement of the F, Cr^{3+}, and Ti^{3+} bands was rather weak. Moreover, it was not possible to detect the etching area by microscopic methods. Figure 1b shows an AFM image of an A-plane sapphire surface after processing in an electron beam, but it is difficult to judge the process of its modification. The surface roughness of sapphire after electron bombardment was less than 0.2 nm. This corresponded to the surface roughness of the initially atomically smooth sapphire substrates. It turned out to be impossible to obtain a high-quality image using electron microscopy (SEM) because of the charging of the dielectric surface of the sapphire.

It is possible, by analogy with [28], that the processes of sapphire etching can be visualized only under long-term electron irradiation (energy 100 keV) for more than 30 min with parallel heating of the sample to at least 1023 K, which is a technologically difficult task.

In the work presented here, it was proposed to accelerate the process of etching sapphire in a beam of bombarding electrons using metal stimulators, particularly gold nanocrystals. The choice of gold was due to the following factors. Gold has weak adhesion to sapphire and, when heated, forms a discrete structure of nano- and microcrystals [29]. The successful implementation of the etching process with metal stimulators could approach a cheaper technology for profiling sapphire. In addition, according to the phase diagram [30], there is a wide range of solid solutions in the gold-rich part of the Al–Au system. There are low-melting phases, such as the rhombohedral phase of Al$_2$Au$_5$; an intermediate- to high-temperature disordered β-phase of the bcc-type, which, upon cooling to 400 °C, transforms into a low-temperature α-phase of AlAu4; a distorted βMn-type phase AlAu4 with a solubility ranging from 80 to 81.2 at.% Au; and an Au fcc-type solid solution with a maximum solid state solubility of 16 at.% Al at 545 °C.

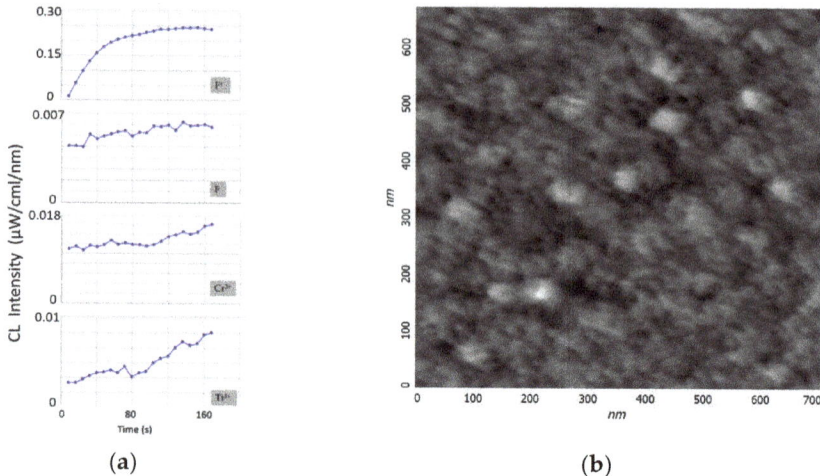

Figure 2. Time dependence (**a**) of the main bands (F$^+$, F, Cr^{3+}, and Ti^{3+}) in the CL spectra for A-plane sapphire (electron energy 40 keV). AFM image (**b**) of a sapphire surface after processing in an electron beam.

In the case of oxygen desorption, the outflow of positively charged aluminum atoms to negatively charged gold crystals can accelerate the decomposition (etching) of the sapphire surface. In addition, there have been recent studies [31] in which gold nanocrystals were used to obtain unstrained GaN films on sapphire.

3.2. A-Plane Etching of Sapphire with Gold Nanocrystals in an Electron Beam with an Energy of 40 keV

In the process of depositing a gold film on A-plane sapphire and heat-treating it, a discrete structure of nanocrystals with lateral sizes from 300 to 700 nm and heights of up to 250 nm was formed (Figure 3a). At the next stage, the sample surface was irradiated with a focused electron beam with an accelerating voltage of 40 kV (beam diameter 0.5 mm) without special heating for 400 s. When examining the surface of the irradiated sample by scanning electron microscopy (SEM), no significant changes were found. Only a slight rounding of gold nanocrystals was observed. A more detailed picture of the effect of electron flows was visualized by scanning using atomic force microscopy (AFM). Etch pits were found in the region of the sapphire surface adjacent to the gold nanocrystals (Figure 3b). The etch pits had an elongated, almost triangular shape. In addition, there were sharp morphological differences (Figure 3c,d) between the areas of treated and untreated electrons. First of all, an increase in the size of nanocrystals was noticeable. A distinctive feature of the etch pits (Figure 4) was that even with different linear dimensions of the pits, they were of the same depth of about 10 nm. This fact indicates the mechanism of layer-by-layer removal of material through intermediate cracking processes [32]. Cracking is caused by local displacement of aluminum atoms and desorption of oxygen atoms, which weaken the interatomic bonds in the lattice. According to our AFM data, the average etching rate of an individual pit was calculated in the order of 10^{-6} μm^3/s.

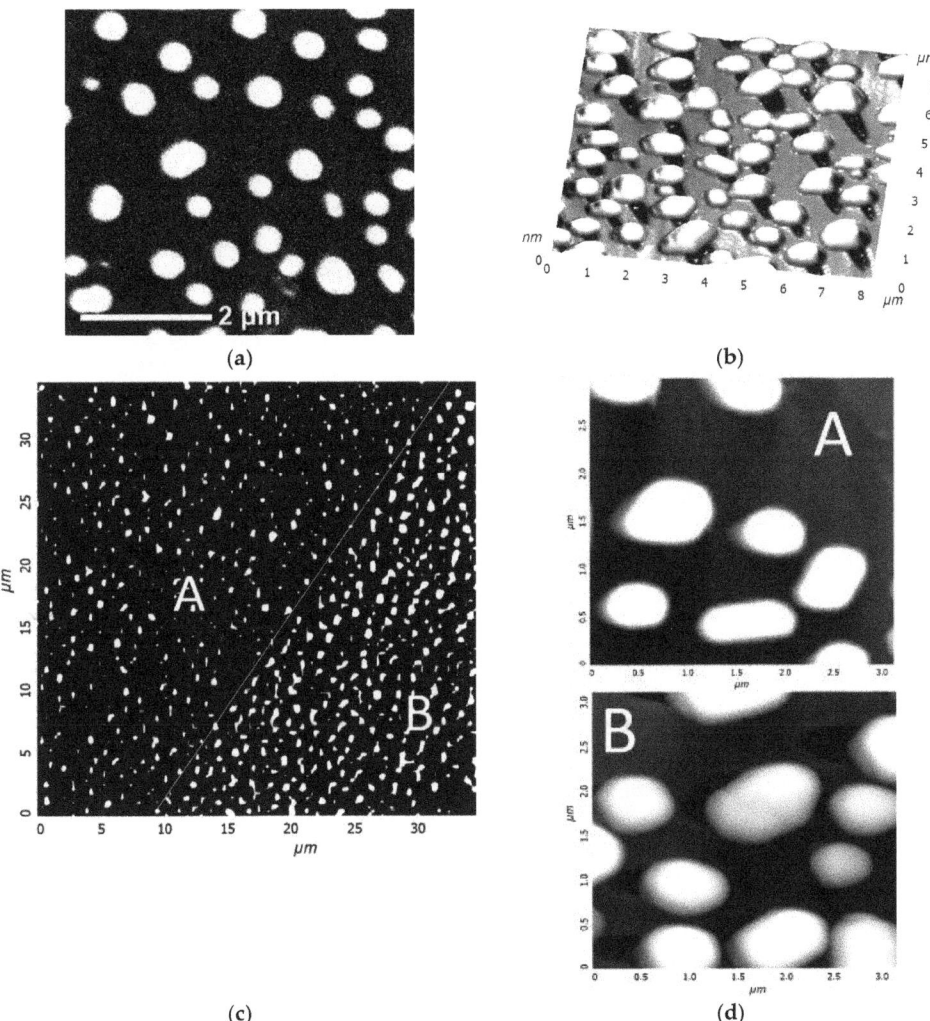

Figure 3. SEM image of A-plane sapphire with gold nanocrystals (**a**). AFM image (3D) of the A-plane of a sapphire with gold nanocrystals after exposure to electrons (**b**). AFM images of adjacent areas A (not treated with electrons) and B (treated with electrons): 35 × 35 μm^2 (**c**), 3 × 3 μm^2 (**d**). The thin white line marks the border of regions A and B.

Interesting results were obtained during the study by probe microscopy in the phase contrast mode, which made it possible to contrastingly display materials of various natures. In Figure 5, one can see the accumulation of material along the borders of a pit with two bases. The high contrast indicates significant differences in the mechanical properties of the cluster materials and the A-plane sapphire surface. Given that the saturated vapor pressure of aluminum is low, oxygen was desorbed during the dissociation of the sapphire surface, and aluminum formed clusters. The accumulations of aluminum had a loose structure, which is the reason for the high contrast of the phase picture. The bond between the aluminum atoms and the sapphire surface was maximum at the boundaries of an etch pit; therefore, aluminum accumulations decorated the profile of the pit.

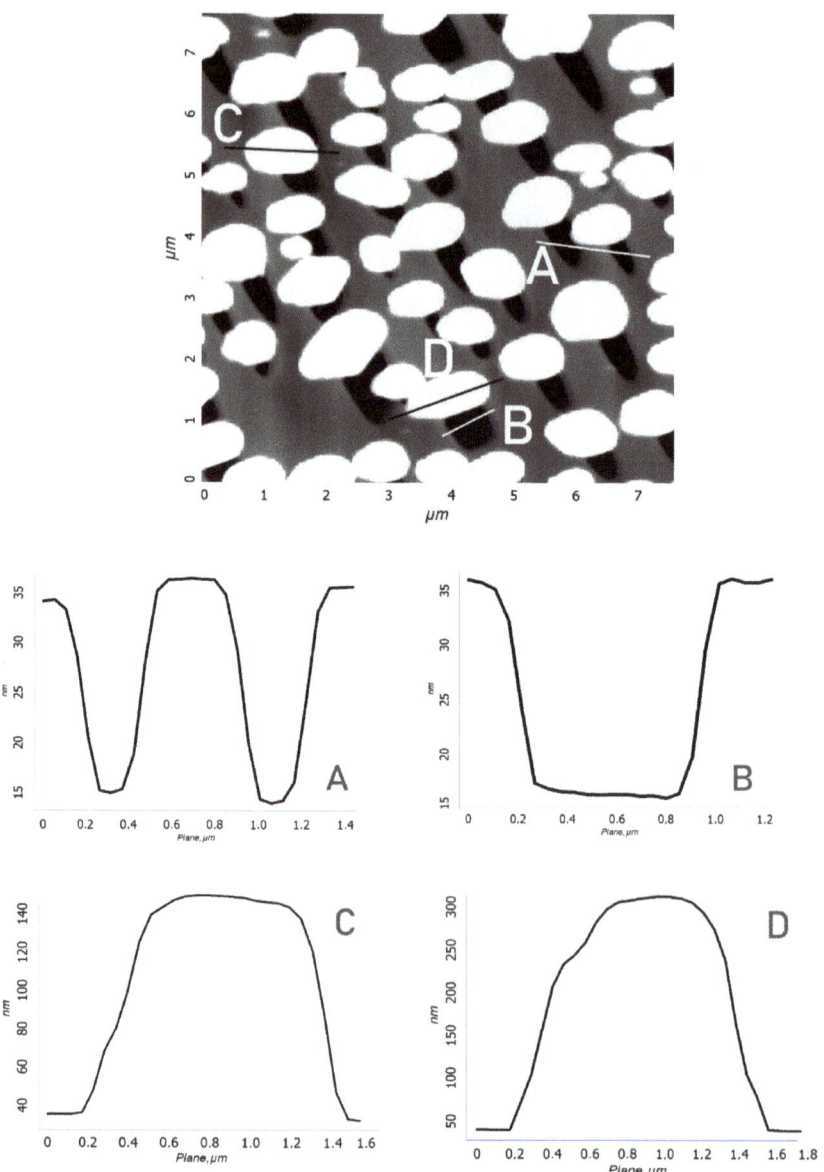

Figure 4. AFM image of an A-plane sapphire with gold nanocrystals after exposure to electrons. Topographic sections of etch pits (**A**,**B**) and of an individual Au nanocrystal (**C**,**D**).

The reduced aluminum atoms were positively charged and diffused to the negatively charged gold islands. The negative charge flowing down from the sapphire surface to the gold islands reduced the effect on the incident electron flux and contributed to the continuity of the etching process. The flow of aluminum atoms to the gold nanocrystal was associated with an increase in their size in the base region (Figure 5b). This was also the reason for the observed rounding of gold nanocrystals when examined by SEM. In the area of the base, a precipitate was formed by compounds of the Al–Au system. According to the phase diagram [30], the most probable compound in the gold-rich region is $AlAu_4$.

However, intense diffusion of the Al and Au components of the solid solution would require heating of this region to ~400–500 °C. It is known that in thin layers, all processes proceed at significantly low temperatures. A refined diagram [33] demonstrates that, in nanometer layers, the phase composition in the gold-rich part suggested several Al_2Au_5 and $AlAu_4$ compounds, and the temperature of active diffusion processes was minimized to 100–200 °C. Considering that the reactions of formation of intermetallic compounds in the Au–Al system are exothermic [33] (they proceed with the release of heat), the Au–Al deposit formation mechanism can be self-sustaining.

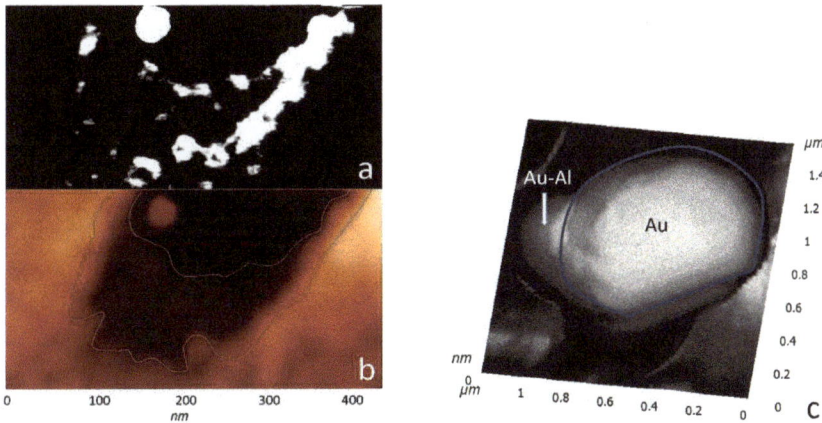

Figure 5. AFM images of an individual etch pit: (**a**)—phase contrast mode; (**b**)—topography mode; (**c**)—3D image of a gold nanocrystal on sapphire after electron impact.

The time dependence of the CL spectra (Figure 6) of an A-plane sapphire sample with gold nanocrystals was studied. For a correct comparison with a pure A-plane sapphire sample without gold, the same beam parameters were used—electron energy of 40 keV and a spot diameter of 0.5 mm. The intensity of the F^+-band of CL, which was proportional to the concentration of oxygen vacancies at the initial stage, also reached the saturation level. However, its absolute value was much lower than in the sample without gold. Gold covered a significant part of the A-plane surface, and active desorption of oxygen proceeded only from the uncovered part. The flow of electrons to gold nanocrystals could also have played an important role. The charge flowed to the gold nanocrystals until their potential was equal to the potential of the cathode. Calculations taking into account the condition of equality of potentials, the electrical capacitance of gold nanocrystals, and the beam parameters showed that their full charging occurred in fractions of a second. This allows us to state that the process of oxygen desorption began from the first seconds, but until the moment (130 s—Figure 6a), it did not have a large-scale character and did not lead to an increase in surface roughness. The intensity of the Cr^{3+} and Ti^{3+} bands at the initial stage did not change and even partially decreased. From 130 s, etching began with the formation of sapphire pits, and the intensity of the Cr^{3+} and Ti^{3+} bands increased. It is interesting that an increase in the etched surface of sapphire was observed up to 200 s, after which the F, Cr^{3+}, and Ti^{3+} bands reached the saturation level, and the rate of increase in the intensity of the F^+-band decreased (Figure 6b). The rate of generation of oxygen vacancies at this stage increased to such an extent that the reverse recombination of vacancies became an important factor. A focused electron beam transferred all active Cr^{3+} and Ti^{3+} ions located in the illumination region into an excited state and, as a result, saturation of the bands was observed.

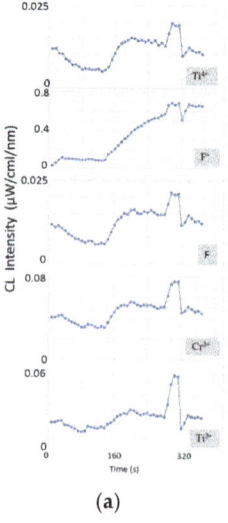
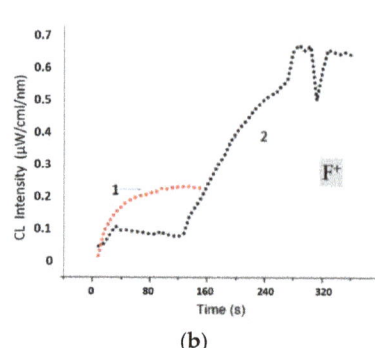

Figure 6. Time dependence of the main bands (Ti^{4+}, F^+, F, Cr^{3+}, Ti^{3+}) in the CL spectra for A-plane sapphire (electron energy 40 keV) with gold nanocrystals (**a**). Comparison (**b**) of the time dependences of the F^+-bands in the CL spectra (electron energy 40 keV) for pure A-plane sapphire (curve 1) and that with gold nanocrystals (curve 2).

Of particular interest is the behavior of the F and Ti^{4+} bands (Figure 6a). The positions of their features on the time dependence coincided with the positions of the other bands, but there were also differences. Note that the F and Ti^{4+} bands indeed luminesced in the complex (their curves were identical). Preliminary annealing of samples in air led to the oxidation of Ti^{3+} centers concentrated in the surface layer to Ti^{4+}. Therefore, the concentration of Ti^{4+} centers dominated at the initial stage. During irradiation, the $Ti^{4+} + e^- \rightarrow (Ti^{3+})^* \rightarrow Ti^{3+}$ + (730 nm) transition took place. This was associated with a decrease in the intensity of the Ti^{4+} band at the initial stage. In addition, the transfer of electrons to F^+-centers from Ti^{4+} centers with the formation of F-centers seemed to be more feasible in sapphire than a direct transition due to electron bombardment. Therefore, a decrease in the intensity of the F–Ti^{4+} complex is observed.

The burst of intensity in all CL bands at 280 s remains incomprehensible (Figure 6). The sharpest surge was observed for the Cr^{3+} and Ti^{3+} bands. Analysis of these bands also demonstrated their maximum broadening at this stage, which indicated the maximum local temperature. It can be assumed that only nearby aluminum atoms, which formed in the etch pit after dissociation of the sapphire surface, diffused to the gold nanocrystal. The most distant ones formed positively charged clusters. Over time, as the clusters grew, their repulsion from each other and from the boundaries of the etch pit, where aluminum accumulations were also observed, increased. At the same time, the force of attraction of aluminum clusters to negatively charged gold nanocrystals increased. As time passed, distant aluminum clusters massively diffused to the gold nanocrystals. Because of multiple reactions of the formation of the compounds Al_2Au_5 and $AlAu_4$, sharp heating was observed in the region of the gold nanocrystal. This contributed to a surge in the intensity of all bands at 280 s and their simultaneous attenuation upon reaching 320 s. It is also important to note a multiple increase in the intensity of CL in the UV region of the spectrum when an A-plane sapphire sample was coated with gold nanocrystals in comparison with a pure A-plane sapphire sample (Figure 6b). This result confirms the importance of the factor of additional heating of the surface due to reactions in the Au–Al system.

In general, the scheme of A-plane etching of sapphire with gold nanocrystals in the process of exposure to electrons can be represented as follows (Figure 7):

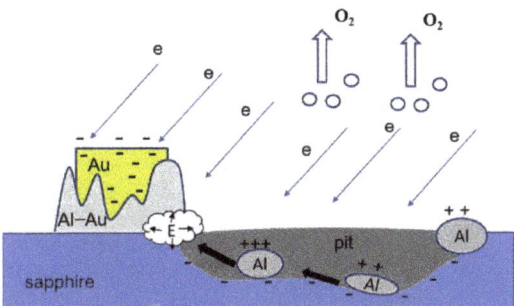

Figure 7. Scheme of A-plane etching of sapphire with gold nanocrystals in the process of exposure to electrons.

3.3. Effects of Inelastic Scattering of Electrons on the Surface of Sapphire

In explaining the results obtained, we took as a basis the mechanism of defect formation caused by the displacement of atoms during elastic collisions with bombarding electrons. Subsequently, because of the weakening of interatomic bonds in the lattice, layer-by-layer removal of material occurs through intermediate cracking processes within the atomic plane. The question arose: are the bombarding electrons capable of leading to displacement of atoms in the sapphire lattice? Calculations [34] showed that the energies required to displace aluminum and oxygen atoms from the sapphire lattice were 18 and 75 eV, respectively. It is possible to consider the kinetic mechanism of electron-stimulated desorption of atoms within which the kinetic energy E_k transferred to an atom of a solid upon collision with an incident electron is determined by the equation:

$$E_k = E_0 \frac{4 m_e m_a}{(m_e + m_a)^2} \cos^2(\varphi) \qquad (2)$$

where m_a and m_e are the masses of the atom and electron, respectively; E_0 is the energy of the incident electron; and φ is the scattering angle. An estimate based on the collision theory shows that when the sapphire surface is bombarded by 40 keV electrons incident at an angle of 45°, the kinetic energies E_k transferred to oxygen and aluminum atoms are ~2.9 eV and 1.7 eV, respectively. Thus, elastically scattered electrons cannot cause the observed etching processes. The thermally activated movement (displacement) of atoms caused by the heating of the sapphire surface under the action of an electron beam can also be considered. Estimates given in [35,36] indicate that the heating of a sapphire surface under the influence of a bombarding electron flow reaches no more than a hundred degrees Celsius. Such heating could have contributed to the acceleration of exothermic reactions in the Al–Au system, but it would have been of negligible effect for the decomposition of the sapphire surface. Presumably, in our case, the effects of inelastic electron scattering were observed. The most probable process is the radiolytic decomposition of sapphire, which is based on the Auger decay effect [37]. More specifically, the process of radiolysis can be represented as follows: as a result of the impact of an external electron, a hole is formed in the inner electron shell Al(2p) of the aluminum ion. After that, one valence electron in O(2p) of the O^{2-} anion jumps into this hole with the release of one or two additional anionic oxygen valence electrons. As a result, the O^{2-} anion changes its charge state and is displaced from the Al_2O_3 lattice. Comparing our data (electron energy 40 keV) with the results of [28], where etching was observed at a higher energy of 100 keV and a duration of 30 min, it can be noted that the cross-section of inelastic electron scattering in a dielectric medium, which causes its radiolysis, decreases with increasing electron energy [38,39]. It can be argued that a decrease in the energy of electrons at energies below 100 keV enhances the processes of radiolysis of the sapphire surface. Heating above 1000 K used in [28] promotes only intense desorption of oxygen and does not directly affect the radiolysis process.

4. Conclusions

Sapphire crystals are widely used in optics and optoelectronics. When using them as optical glasses of devices operated in a wide variety of environments, including spacecraft, resistance to the effects of various types of ionizing radiation in vacuum conditions and flows of charged particles of various energies is required. On the other hand, technological optimization of the stages of pregrowth preparation of the surface of sapphire substrates (polishing, microstructuring, profiling) that are important in the formation of light-emitting devices is required. In the present work, the results of studying the processes of the action of an electron flow under vacuum conditions on the surface of sapphire substrates of various orientations are presented. The effect of etching a sapphire surface at room temperature in a vacuum by an electron beam with an energy of 70 keV or less was discovered. It was shown that the highest etching rate was observed for the A-plane of sapphire, while the lowest was observed for the C-plane of sapphire. A technique for metal-stimulated etching of a sapphire surface is proposed in which the etching rate increased many times over. The stimulation process was based on exothermic reactions of formation of intermetallic compounds with aluminum reduction on the sapphire surface. For in situ study of the process of etching a sapphire surface, the technique of cathodoluminescence analysis was used for the first time. Possible mechanisms of sapphire etching by a beam of bombarding electrons were considered. An estimate based on the collision theory showed that the bombardment of the sapphire surface by electrons with an energy of 70 keV or less could not have caused the observed etching processes. A sapphire surface etching mechanism based on the effects of inelastic electron scattering is proposed. The most probable process was the process of radiolytic decomposition of sapphire (Auger decay effect). The results obtained in this work could be important in solving the problem of resistance to external radiation effects of sapphire windows of spacecraft. In addition, the proposed method of metal-stimulated etching of a sapphire surface could be widely used in PSS technology and further forming low-dislocation light-emitting structures on patterned sapphire substrate.

Author Contributions: A.M. and V.K. carried out the experiments; participated in writing, reviewing, and editing documents; provided financial and technical support; and supervised the entire research process. All authors have read and agreed to the published version of the manuscript.

Funding: This research was performed in the frame of state assignments of the Ministry of Science and Higher Education of the Russian Federation for FSRC "Crystallography and Photonics" RAS and partially funded by RFBR (research project № 20-21-00068 ROSATOM).

Institutional Review Board Statement: Not applicable.

Informed Consent Statement: Not applicable.

Data Availability Statement: Not applicable.

Acknowledgments: The authors are grateful to A.M. Ismailov for their help in preparing samples. The authors are grateful for technical support from the collective use centers of FSRC "Crystallography and Photonics" RAS.

Conflicts of Interest: The authors declare no conflict of interest. The funders had no role in the design of the study; in the collection, analyses, or interpretation of data; in the writing of the manuscript; or in the decision to publish the results.

References

1. Wang, G.; Zuo, H.; Zhang, H.; Wu, Q.; Zhang, M.; He, X.; Hu, Z.; Zhu, L. Preparation, quality characterization, service performance evaluation and its modification of sapphire crystal for optical window and dome application. *Mater. Des.* **2009**, *31*, 706–711. [CrossRef]

2. Nakamura, S.; Mukai, T.; Senoh, M. Candela-class high-brightness InGaN/AlGaN double-heterostructure blue-light-emitting diodes. *Appl. Phys. Lett.* **1994**, *64*, 1687–1689. [CrossRef]
3. Du, C.; Ma, Z.; Zhou, J.; Lu, T.; Jiang, Y.; Jia, H.; Liu, W.; Chen, H. Modulating emission intensity of GaN-based green light emitting diodes on c-plane sapphire. *Appl. Phys. Lett.* **2014**, *104*, 151102. [CrossRef]
4. Amano, H.; Akasaki, I.; Hiramatsu, K.; Koide, N.; Sawaki, N. Effects of the buffer layer in metalorganic vapour phase epitaxy of GaN on sapphire substrate. *Thin Solid Films* **1988**, *163*, 415–420. [CrossRef]
5. Huang, X.-H.; Liu, J.-P.; Fan, Y.-M.; Kong, J.-J.; Yang, H.; Wang, H.-B. Improving InGaN-LED performance by optimizing the patterned sapphire substrate shape. *Chin. Phys. B* **2012**, *21*, 037105. [CrossRef]
6. Ashby, C.I.H.; Mitchell, C.C.; Han, J.; Missert, N.A.; Provencio, P.P.; Follstaedt, D.M.; Peake, G.M.; Griego, L. Low-dislocation-density GaN from a single growth on a textured substrate. *Appl. Phys. Lett.* **2000**, *77*, 3233. [CrossRef]
7. Takeuchi, T.; Sota, S.; Katsuragawa, M.; Komori, M.; Takeuchi, H.; Amano, H.; Akasaki, I. Quantum-Confined Stark Effect due to Piezoelectric Fields in GaInN Strained Quantum Wells. *Jpn. J. Appl. Phys.* **1997**, *36*, L382–L385. [CrossRef]
8. Takeuchi, T.; Wetzel, C.; Yamaguchi, S.; Sakai, H.; Amano, H.; Akasaki, I.; Kaneko, Y.; Nakagawa, S.; Yamaoka, Y.; Yamada, N. Determination of piezoelectric fields in strained GaInN quantum wells using the quantum-confined Stark effect. *Appl. Phys. Lett.* **1998**, *73*, 1691–1693. [CrossRef]
9. Ni, X.; Özgür, U.; Baski, A.; Morkoç, H.; Zhou, L.; Smith, D.J.; Tran, C.A. Epitaxial lateral overgrowth of (11$\bar{2}$2) semipolar GaN on (1$\bar{1}$00) m-plane sapphire by metalorganic chemical vapor deposition. *Appl. Phys. Lett.* **2007**, *90*, 182109. [CrossRef]
10. Anuar, A.; Makinudin, A.H.A.; Al-Zuhairi, O.; Chanlek, N.; Abu Bakar, A.S.; Supangat, A. Growth of semi-polar (11$\bar{2}$2) GaN on m-plane sapphire via In-Situ Multiple Ammonia Treatment (I-SMAT) method. *Vacuum* **2020**, *174*, 109208. [CrossRef]
11. Saito, Y.; Okuno, K.; Boyama, S.; Nakada, N.; Nitta, S.; Ushida, Y.; Shibata, N. m-Plane GaInN Light Emitting Diodes Grown on Patterneda-Plane Sapphire Substrates. *Appl. Phys. Express* **2009**, *2*, 41001. [CrossRef]
12. Seo, Y.G.; Baik, K.H.; Song, H.; Son, J.-S.; Oh, K.; Hwang, S.-M. Orange a-plane InGaN/GaN light-emitting diodes grown on r-plane sapphire substrates. *Opt. Express* **2011**, *19*, 12919–12924. [CrossRef] [PubMed]
13. Savvides, N. Surface microroughness of ion-beam etched optical surfaces. *J. Appl. Phys.* **2005**, *97*, 53517. [CrossRef]
14. Wen, Q.; Wei, X.; Jiang, F.; Lu, J.; Xu, X. Focused Ion Beam Milling of Single-Crystal Sapphire with A-, C-, and M-Orientations. *Materials* **2020**, *13*, 2871. [CrossRef]
15. Canut, B.; Ramos, S.M.M.; Thevenard, P.; Moncoffre, N.; Benyagoub, A.; Marest, G.; Meftah, A.; Toulemonde, M.; Studer, F. High energy heavy ion irradiatiom effects in α-Al2O3. *Nucl. Instrum. Methods B* **1993**, *80–81*, 1114–1118. [CrossRef]
16. Sporea, D.; Sporea, A. Radiation effects in sapphire optical fibers. *Phys. Status Solidi (C)* **2007**, *4*, 1356–1359. [CrossRef]
17. Gangnaik, A.S.; Georgiev, Y.M.; Holmes, J.D. New Generation Electron Beam Resists: A Review. *Chem. Mater.* **2017**, *29*, 1898–1917. [CrossRef]
18. Martins, M.; Silva, T. Electron accelerators: History, applications, and perspectives. *Radiat. Phys. Chem.* **2014**, *95*, 78–85. [CrossRef]
19. Thune, E.; Fakih, A.; Matringe, C.; Babonneau, D.; Guinebretière, R.J. Understanding of one dimensional ordering mechanisms at the (001) sapphire vicinal surface. *Appl. Phys.* **2017**, *21*, 15301–15302. [CrossRef]
20. Thapa, J.; Liu, B.; Woodruff, S.D.; Chorpening, B.T.; Buric, M.P. Raman scattering in single-crystal sapphire at elevated temperatures. *Appl. Opt.* **2017**, *56*, 8598. [CrossRef]
21. Toshima, R.; Miyamaru, H.; Asahara, J.; Murasawa, T.; Takahashi, A. Ion-induced Luminescence of Alumina with Time-resolved Spectroscopy. *J. Nucl. Sci. Technol.* **2002**, *39*, 15–18. [CrossRef]
22. Molnár, G.; Benabdesselam, M.; Borossay, J.; Lapraz, D.; Iacconi, P.; Kortov, V.; Surdo, A. Photoluminescence and thermoluminescence of titanium ions in sapphire crystals. *Radiat. Meas.* **2001**, *33*, 663. [CrossRef]
23. Costantini, J.-M.; Watanabe, Y.; Yasuda, K.; Fasoli, M. Cathodo-luminescence of color centers induced in sapphire and yttria-stabilized zirconia by high-energy electrons. *J. Appl. Phys.* **2017**, *121*, 153101. [CrossRef]
24. Gibson, U.J.; Chernuschenko, M. Ruby films as surface temperature and pressure sensors. *Opt. Express* **1999**, *4*, 443–448. [CrossRef] [PubMed]
25. Ghamnia, M.; Jardin, C.; Bouslama, M. Luminescent centres F and F+ in α-alumina detected by cathodoluminescence technique. *J. Electron Spectrosc. Relat. Phenom.* **2003**, *133*, 55–63. [CrossRef]
26. Cao, L.; Zhang, X.; Yuan, J.; Guo, L.; Hong, T.; Hang, W.; Ma, Y. Study on the Influence of Sapphire Crystal Orientation on Its Chemical Mechanical Polishing. *Appl. Sci.* **2020**, *10*, 8065. [CrossRef]
27. Mikhailik, V.; Di Stefano, P.C.F.; Henry, S.; Kraus, H.; Lynch, A.; Tsybulskyi, V.; Verdier, M.A. Studies of concentration dependences in the luminescence of Ti-doped Al_2O_3. *J. Appl. Phys.* **2011**, *109*, 53116. [CrossRef]
28. Chen, C.; Furusho, H.; Mori, H. Effects of temperature and electron energy on the electron-irradiation-induced decomposition of sapphire. *Philos. Mag. Lett.* **2010**, *90*, 715–721. [CrossRef]
29. Pandey, P.; Sui, M.; Li, M.-Y.; Zhang, Q.; Kim, E.-S.; Lee, J. Shape transformation of self-assembled Au nanoparticles by the systematic control of deposition amount on sapphire (0001). *RSC Adv.* **2015**, *5*, 66212–66220. [CrossRef]
30. Massalski, T.B.; Okamoto, H.; Subramanian, P.R.; Kacprzak, L. (Eds.) Metals Park. In *Binary Alloy Phase Diagrams*; American Society for Metals: Novelty, OH, USA, 1986; Volume 1, 90p.
31. Li, P.; Xiong, T.; Wang, L.; Sun, S.; Chen, C. Facile Au-assisted epitaxy of nearly strain-free GaN films on sapphire substrates. *RSC Adv.* **2020**, *10*, 2096–2103. [CrossRef]

32. Kim, S.H.; Sohn, I.-B.; Jeong, S. Ablation characteristics of aluminum oxide and nitride ceramics during femtosecond laser micromachining. *Appl. Surf. Sci.* **2009**, *255*, 9717–9720. [CrossRef]
33. Murray, J.L.; Okamoto, H.; Massalski, T.B. The Al−Au (Aluminum-gold) system. *Bull. Alloy Phase Diagr.* **1987**, *8*, 20–30. [CrossRef]
34. Pells, G.; Phillips, D. Radiation damage of α-Al2O3 in the HVEM: I. Temperature dependence of the displacement threshold. *J. Nucl. Mater.* **1979**, *80*, 207–214. [CrossRef]
35. Gossink, R.; Van Doveren, H.; Verhoeven, J. Decrease of the alkali signal during auger analysis of glasses. *J. Non-Cryst. Solids* **1980**, *37*, 111–124. [CrossRef]
36. Milakhin, D.; Malin, T.; Mansurov, V.; Galitsyn, Y.; Zhuravlev, K. Electron-Stimulated Aluminum Nitride Crystalline Phase Formation on the Sapphire Surface. *Phys. Status Solidi (B)* **2019**, *256*, 1800516. [CrossRef]
37. Knotek, M.L.; Feibelman, P. Ion Desorption by Core-Hole Auger Decay. *Phys. Rev. Lett.* **1978**, *40*, 964–967. [CrossRef]
38. Egerton, R. Mechanisms of radiation damage in beam-sensitive specimens, for TEM accelerating voltages between 10 and 300 kV. *Microsc. Res. Technol.* **2012**, *75*, 1550–1556. [CrossRef]
39. Hooley, R.; Brown, A.; Brydson, R. Factors affecting electron beam damage in calcite nanoparticles. *Micron* **2019**, *120*, 25–34. [CrossRef]

Article

Influence of Beam Figure on Porosity of Electron Beam Welded Thin-Walled Aluminum Plates

Matthias Moschinger [1,*], Florian Mittermayr [2] and Norbert Enzinger [1]

[1] Institute of Materials Science, Joining and Forming, Graz University of Technology, 8010 Graz, Austria; norbert.enzinger@tugraz.at

[2] Institute of Technology and Testing of Building Materials, Graz University of Technology, 8010 Graz, Austria; f.mittermayr@tugraz.at

* Correspondence: matthias.moschinger@tugraz.at

Abstract: Welded aluminum components in the aerospace industry are subject to more stringent safety regulations than in other industries. Electron beam welding as a highly precise process fulfills this requirement. The welding of aluminum poses a challenge due to its high tendency to pore formation. To gain a better understanding of pore formation during the process, 1.5 mm thick aluminum AW6082 plates were welded using specially devised beam figures in different configurations. The obtained welds were examined with radiographic testing to evaluate the size, distribution, and the number of pores. Cross-sections of the welds were investigated with light microscopy and an electron probe microanalyzer to decipher the potential mechanisms that led to porosity. The examined welds showed that the porosity is influenced in various ways by the used figures, but it cannot be completely avoided. Chemical and microstructural analyzes have revealed that the main mechanism for pore formation was the evaporation of the alloying elements Mg and Zn. This study demonstrates that the number of pores can be reduced and their size can be minimized using a proper beam figure and energy distribution.

Keywords: electron beam welding; aluminum 6082; porosity; beam figure

Citation: Moschinger, M.; Mittermayr, F.; Enzinger, N. Influence of Beam Figure on Porosity of Electron Beam Welded Thin-Walled Aluminum Plates. *Materials* **2022**, *15*, 3519. https://doi.org/10.3390/ma15103519

Academic Editors: Katia Vutova and Pavel Diko

Received: 7 March 2022
Accepted: 11 May 2022
Published: 13 May 2022

Publisher's Note: MDPI stays neutral with regard to jurisdictional claims in published maps and institutional affiliations.

Copyright: © 2022 by the authors. Licensee MDPI, Basel, Switzerland. This article is an open access article distributed under the terms and conditions of the Creative Commons Attribution (CC BY) license (https://creativecommons.org/licenses/by/4.0/).

1. Introduction

Saving fuel and increasing payload are two of the biggest goals in aviation and aerospace. Manufacturing parts in lightweight construction is one way to support this goal. Low density and high strength make aluminum an excellent construction material for such lightweight components that are already being used in this sector. However, the joining of aluminum materials is often challenging [1,2]. Especially in the aerospace industry, the limitations regarding defect sizes and processes are obviously very strict [3].

Porosity in aluminum welding is a well-known and critical phenomenon. According to the literature, the formation of porosity in the weld can have different reasons, such as the processes, but also the base materials' chemical composition and/or filler materials play an important role. For instance, the introduction of hydrogen into the molten pool is of great importance, as it is highly soluble in liquid aluminum [4].

The alloy burn-off of certain elements is a further mechanism causing pore formation. Zhang et al. described the effect on electron beam welded (EBW) AA6062 [5]. Zhou et al. investigated this elemental burn-off effect in detail in their article on plasma welding of an AW5052 [6]. The welded Al-alloys in both of these studies displayed the burn-off mechanism for Mg.

The formation of a gaseous phase from aluminum and its oxide is another pore-forming mechanism that is related to the chemistry of the material as described by Fujii et al. [7]. According to the authors, it can only occur in a high vacuum, such as during the EBW process due to the low partial pressure of the oxide. Nogi et al. [8] showed that the oxide thickness correlated with the porosity during EBW. This is also a conceivable mechanism for porosity, but it is difficult to detect.

There is also the possibility of pore formation solely from the process side. During EBW, a keyhole can form due to the high energy density, which causes complex flow processes in the weld pool [9]. However, these processes have not yet been fully understood as the experimental detection of such flows is a challenge [10,11]. Chen et al. suggested that these flows cause the movement of bubbles within the melt pool. If the bubbles come close to the solidification front, they can be "frozen" and thus remain as pores in the weld.

The beam deflection also leads to a manipulation of the molten pool and thus influences the fluid flow. At present, the influences on the melt pool of beam oscillation are rarely investigated. Wang et al. [12] studied oscillation movements describing the behavior of longitudinal, transverse, and circular oscillation on the molten bath and its influence on solidification patterns. In the case of superimposed oscillations, the shape of the "eight" appears in the literature, in addition to the circle; both represent so-called Lissajous figures. Kabasakaloglu et al. [13] and Chen et al. [14] investigated the influences of the beam deflection on the porosity. In both studies the figure of the "eight", either standing (8) or lying (∞) was selected.

The latter mentioned studies were all performed on laser welded aluminum under atmospheric conditions. Since the laser beam is always limited in its deflection speed (inertia due to robot mass, mirror kinematics), only beam deflections in the lower speed spectrum can be achieved. On the contrary, in the case of EBW, the electron beam is moved using magnetic fields, and thus a significantly higher deflection speed can be realized. Thus, the current study deals with the influences of electron beam deflection on porosity. For this purpose, new beam figures were devised for the welding process. The aims of this study are (i) to counteract the formation of gas bubbles by optimizing the beam figures, (ii) to facilitate the escape of potential bubbles from the melt pool by extending it, and (iii) to investigate the underlying mechanisms that lead to porosity during EBW on AW6082.

2. Materials and Methods

2.1. Material

The aluminum used in this work is the commercial alloy AW6082-T6, i.e., in the aged condition [15]. The chemical limits of the composition of the material used are shown in Table 1. The used specimens were plates with a length of 300 mm and a width of 60 mm. The plates have a thickness of 1.5 mm, but they have a material accumulation at their butt edge to ensure sufficient material supply during the welding process. This is necessary due to the required seam elevation and root penetration. The joint geometry is produced by milling. This processing also has the purpose of removing the existing oxide layer. To prevent a renewed formation of this oxide layer as far as possible, the milled plates were packed airtight directly after processing.

Table 1. Limits of the chemical composition of the aluminum alloy AW6082 [16]. Copyright 2020 Ansys Granata Edupack.

Al	Cr	Cu	Fe	Mg	Mn	Si	Ti	Zn	Others
95.2	0	0	0	0.6	0.4	0.7	0	0	0
98.3	0.25	0.1	0.5	1.2	1	1.3	0.1	0.2	0.15

2.2. Welding Process and Beam Figures

The EBW machine used was a chamber system of the type EBG 45-150K14 from Pro-beam GmbH & Co. KGaA. Since EBW allows a wide range of welding parameter combinations, welding parameters were determined by preliminary tests. For this purpose, blind welds were first welded and the parameters obtained were adapted to the joint geometry. The majority of the weld parameters were kept constant over all tests. The EB was focused on the plate surface. Welding was performed at a welding speed of 20 mm/s.

The chamber pressure during welding was below 5×10^{-4} mbar. The beam oscillation was achieved using beam figures with a frequency of 500 Hz. Welding was performed with a beam voltage of 80 kV. The beam power could not be kept constant over all tests, because the beam figures had different shapes and different energy distributions. Therefore, the beam current varied in the range of 9.7–13 mA.

$$t \rightarrow \begin{pmatrix} A_x \sin(\omega_1 t + \varphi_1) \\ A_y \sin(\omega_2 t + \varphi_2) \end{pmatrix}, \ t \in [0, \infty) \qquad (1)$$

The beam oscillated using superimposed oscillations in x and y-directions which were parameterized according to Equation (1). A frequency ratio $\omega_1:\omega_2$ of 2:3 and a phase shift of $\Delta\varphi = \varphi_1 - \varphi_2 = 0°$ resulted in a figure as shown in Figure 1. This Lissajous figure represents the reference figure R with an overall width of 0.45 mm and a total length of 1.35 mm. All other figures were derived from the reference figure: Several variations of the figure were created which differ in terms of their energy distribution and dimensions. To evaluate the overall energy distribution, the trajectory of the figure was analyzed, as shown in Figure 2. For this purpose, the points describing the beam figure were considered to have a constant energy input and were integrated over the welding path.

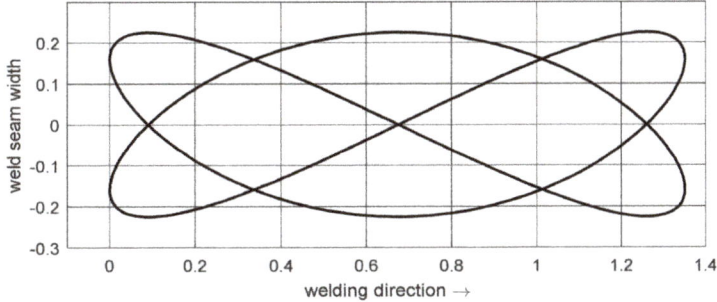

Figure 1. Reference figure R which is the basis for the devised figures.

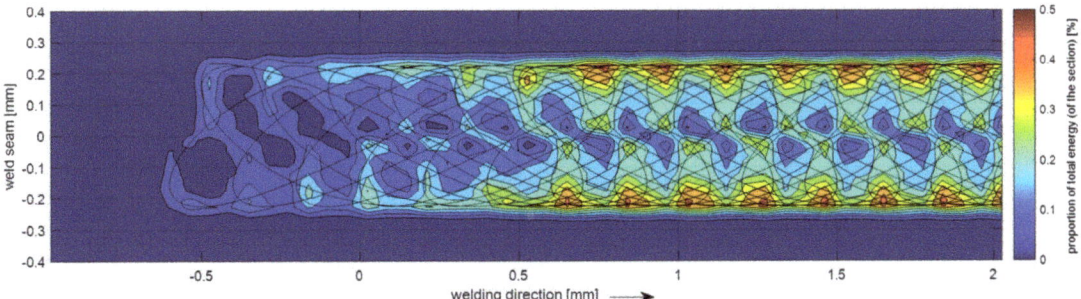

Figure 2. Relative energy distribution of the moving reference figure R starting on the left and reaching steady state on the right side.

In total 46 welding experiments were performed, 16 in the first set and 30 in the second set of experiments. In the first set, four figures were welded in four different configurations (the configurations are shown in Figure A1 in Appendix A). The resulting 16 welds were examined and based on these preliminary results a second experimental set was designed and consequently welded (marked in red and labeled in Table A1).

For this second experiment set, five different figure configurations were selected from the first experiment set (see Tables 2 and A1). At least one configuration of each figure was repeated, the decision was based on porosity and weld appearance (root and seam).

The figure configurations of series A-C & E were chosen for their low porosity appearance. Figure configuration D was chosen because of the high pore count. It was investigated whether high porosity can be reproduced. The set was completed by the reference figure R. Every configuration was repeated five times, to investigate the reproducibility of the results. This resulted in a total weld length of 1.3 m minus the start and end sections of the welded plates, so only the stationary part of the weld was investigated.

Table 2. Welding figures and their parameters for the second series of experiments. The energy distribution was symmetrical across the welding direction. All figures contained 1000 dots.

Series	Figure Welding Direction →	Energy Distribution	Length [mm]	Width [mm]
A		quadratic	1.35	0.45
B		linear	3	0.45
C		centered	1.35	0.45
D		constant	1.35	0.25
E		Right Figure 75% Left Figure 25%	3	0.45

- The series A and B were derived from the same basic figure which has an energy reduction from the front to the back, which was either linear or quadratic. In addition to the energy distribution, the length of the figure was also changed. Thus, A and B represent two figure configurations with different lengths and energy distributions as shown in Table 2.
- For the basic figure of the series C, the energy was shifted to the weld centerline, since the analysis of the energy distribution of the reference figure R showed that most of the energy accumulated on the figure edge (see Figure 2). The figure was varied in length and the gradient of the energy distribution, the series C configuration was finally chosen for this figure type.
- Series D was developed based on the reference figure R with constant energy distribution. However, the width was decreased to make the weld narrower and thus give the pores less space to form. The selected figure width for series D was 0.25 mm (see Figure 2).
- The basis for the series E was a multi-bead technique as known from the literature [17]. Here, two figures were applied sequentially, the first figure ensuring the full penetration welding; the second figure is a post-heating of the molten pool. The distance between the individual figures (center to center) of 1.65 mm was used, which leads to a total length of 3 mm (front of the right figure to back of the left figure).

Due to the low beam current (9.7–13 mA) of the EB and thus the low beam power the EB was susceptible to disturbances. Such a disturbance can be, e.g., the rising metal vapor during the process. To counteract metal vapor disturbances, the beam was tilted by 5.7°, see Figure 3a. The tilting of the EB distorted the beam figure in the horizontal plane. To correct this, the plates were also tilted at this angle using a swiveling clamping device (Figure 3b).

Before mounting, the plates were unpacked and cleaned with isopropanol. To minimize oxidation as much as possible, the welding chamber was closed and evacuated about 3 min after unpacking the sample.

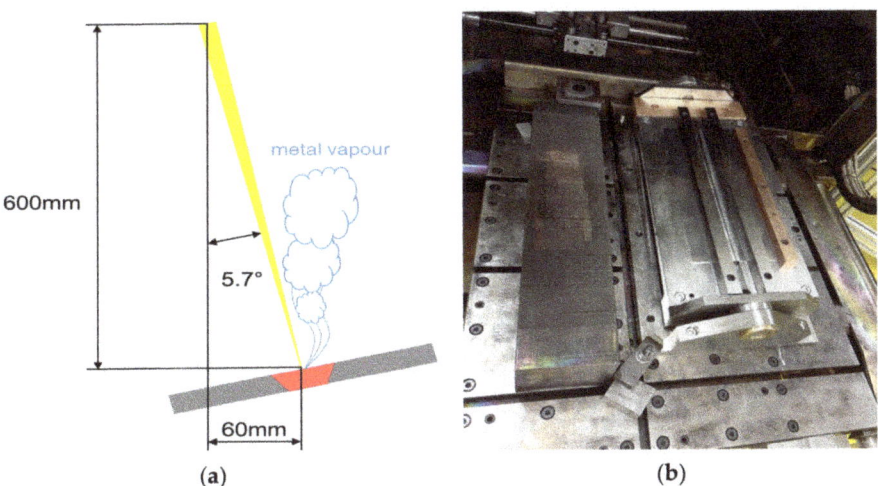

Figure 3. (a) Schematic of the tilted beam; (b) Pivoting clamping device with welded plates.

Before actual welding, the plates were tack welded using the reference figure R with a current of 6 mA. The tack weld was welded over a length of 290 mm, of which 15 mm was each slope in and slope out. This slope in/out was not considered for the evaluation, and therefore, arbitrarily selected; it was intended to provide a smooth entry and exit area for the weld. After tack welding, a waiting time of 5 min was considered before the actual welding process was carried out, also with the slope in/out. After welding, a 2 min waiting time was defined before the chamber was flooded with air, and the plates were removed from the machine.

2.3. Investigations

From the first experimental set, samples were taken 50 mm from the beginning of the weld and analyzed metallographically. The longer parts of the weld, about 220 mm each, were examined for pores using radiographic testing. The second experimental set was only investigated radiographically.

To investigate the microstructure, cross-sections were prepared using BARKER etching [18]. The etching was carried out at 25 V with a flow rate of 15 L/min over a time of 120 s. A light microscope (Zeiss Axio Observer Inverted) was used to view the cross-sections. Images were taken using polarized light incident at an angle of 3.5° on the sample surface.

To detect the porosity in the welds, radiographic testing was used. The entire length of the weld was irradiated without further sample preparation. This was conducted with the Yxlon Cheetah system, which was operated with a microfocus X-ray tube. The working voltage of the tube was 120 kV at 90 mA current. The radiography resulted in 15 X-ray images covering the entire length of the weld. An example image is shown in Figure 4a. First, the images were automatically analyzed using the Yxlon Fgui software with 14× magnification. Then the X-ray images were manually re-inspected and the data were corrected as the software also identified non-porosity related objects (e.g., weld butts) in the areas of weld start/end see Figure 4b.

The chemical composition combined with the microstructure of welds and the base material was analyzed using an EPMA (Electron Probe Microanalyzer, JEOL JXA8530F Plus Hyper Probe) equipped with a Schottky field emission gun on the welded reference sample. Therefore, several cross-sections through welds (with and without visible pores) were pre-selected from the light microscopic investigations, polished and C-coated. Images were recorded in the backscattered electron (BSE) and in the secondary electron (SE) modes at 15 kV acceleration voltage and 20 nA beam current. Semi-quantitative elemental distribution images of aluminum, magnesium, silicon, manganese and iron were acquired from selected areas with the same acceleration voltage, a beam current of 50 nA, a dwell time of 15 mS and a focused beam. The dimension of the mapping showing both sides of the fusion line (Figure 7 and Figure A3) is 1000 × 350 px and was recorded with a 1 μm step size, while mappings with a large pore (Figure 8 and Figure A4) are 1000 × 1000 px and were recorded with a 0.5 μm step size. The quantification of the individual element mappings in wt.% was performed against certified pure metal standards. The quantitative single spot analyses were performed on the Al-phase both in the welds and base material. The Si-eutectic and other phases were not investigated. The analytical conditions were as follows: 15 kV accelerating voltage, 20 nA beam current and a defocused beam with a diameter of 40 μm. The full quantitative chemical analyses were also standardized against certified pure metal standards and included the following elements: Al, Si, Mg, Fe, Cu, Zn, Mn and Cr. The counting times on peak and on both sides on the background positions were 20 and 10 s, respectively. The detection limits were 117 ppm for Al, 89 ppm for Si, 140 ppm for Mg, 238 ppm for Fe, 294 ppm for Cu, 639 ppm for Zn, 216 ppm for Mn and 246 ppm for Cr. Results presented in Table 3 show integrated averages in wt.% from the highlighted areas in Figure 7 and Figure A4. For the single spot analyses, mean values for the weld areas (n = 27) and the base material (n = 18) are shown. The results of the individual single spot analyses are placed in the Appendix A (Table A1).

Figure 4. (**a**) example of an X-ray image of the test series D with different sized pores; (**b**) misinterpreted weld butt at the end of the weld.

Table 3. EPMA results from the single spot analyses and the integrated mapping areas; SD: standard deviation; bld.: below detection limit.

Quant Results	Single Spot Analyses	Al	Si	Mg	Fe	Mn	Cr	Cu	Zn	Total
	[n]	[wt.%]	[wt.%]	[wt.%]	[wt.%]	[wt.%]	[wt.%]	[wt.%]	[wt.%]	[wt.%]
Avg. base material	18	96.21	1.06	0.85	0.42	0.64	0.13	0.10	0.13	99.54
±(SD)		0.99	0.30	0.16	0.42	0.24	0.04	0.02	0.02	0.30
Avg. weld	27	96.08	1.27	0.49	0.49	0.68	0.14	0.10	bld.	99.24
±(SD)		0.50	0.28	0.05	0.09	0.04	0.01	0.02	-	0.29
Semi-Quant Results	Integrated Mapping Area	Al	Si	Mg	Fe	Mn	Cr	Cu	Zn	Total
	[mm^2]	[wt.%]	[wt.%]	[wt.%]	[wt.%]	[wt.%]	[wt.%]	[wt.%]	[wt.%]	[wt.%]
Avg. base material	38.62 (Figure 7)	96.45	1.12	0.93	0.55	0.56	not analyzed			99.61
Avg. weld	30.19 (Figure 7)	97.02	1.21	0.56	0.63	0.58	not analyzed			99.99
Avg. weld	0.44 (Figure 8 and Figure A1)	96.82	1.18	0.52	0.61	0.58	not analyzed			99.71

3. Results

3.1. Metallography

The cross-sections showed little difference in shape and microstructure (Figure A2). In Figure 5a, the cross-section of reference R is shown. The microstructural investigation valid for all welding tests clearly revealed two different areas, a dendritic area starting from the fusion line and a globular area in the center of the weld. The globular region represents equiaxed grains, as shown by a longitudinal section in the middle of the weld (Figure 5b).

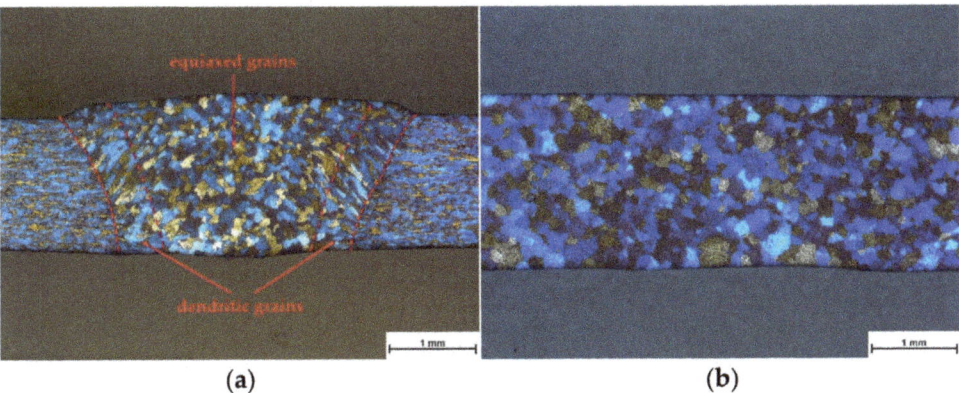

Figure 5. (**a**) cross-section of the reference figure R; (**b**) longitudinal section of the reference figure.

3.2. Porosity

The results shown in Figure 6 are from the second set of experiments. The diagram shows, on the one hand, the number of pores and on the other hand a boxplot diagram [19] of the pore diameters. It can be seen that there are significant differences between the tests, in terms of the number of pores, and the average pore diameter. For instance, series B has the lowest number of pores (7), but the largest average pore diameter (0.80 mm). On the contrary, series A has the most pores (54), but the smallest average pore diameter (0.20 mm) and a low deviation around the average pore size. Series D recorded a high number of pores (52), the second largest average pore size (0.58 mm) of all tests, and by far the largest deviation around the average. Series C emerged with a relatively low number of pores (18), while also having a relatively low average pore diameter of 0.22 mm. The reference test R and series E had similar numbers of pores (R: 35; E: 30). However, these two tests had slightly different average pore sizes of 0.30 and 0.24 mm.

3.3. Chemical Composition and Microstructure Revealed by EPMA

The following investigations were performed on samples that were welded with the reference beam figure R. The intense heating as a consequence of the EBW process clearly caused (micro)structural and chemical modifications to the Al-alloy. Figure 7, for example, shows an elemental mapping of magnesium. The fusion line (indicated in red) is well visible and it marks a distinct structural and chemical border between the base material (right side) and the weld area (the left side of the image is the center of the weld). The Mg distribution shows a more homogeneous distribution and the presence of relatively large Mg and Si containing phases in the base material, which is due to the aging process of the alloy. On the contrary, in the weld, the Mg is depleted and much less homogeneously distributed. The quantification of the map and the results from the single spot analyses (see Tables 3 and A1) both indicate that the resolidified weld contained approximately 40 wt.% less Mg compared to the base material. For all the other elements, the analyses of the base metal match well with the boundary values given for AW6082 (see Table 1). Besides the Mg depletion, a clear reduction of Zn was recorded. In the base material 0.13 ± 0.02 wt.% Zn was found while in the weld the Zn-content was below the detection level of ≈600 ppm.

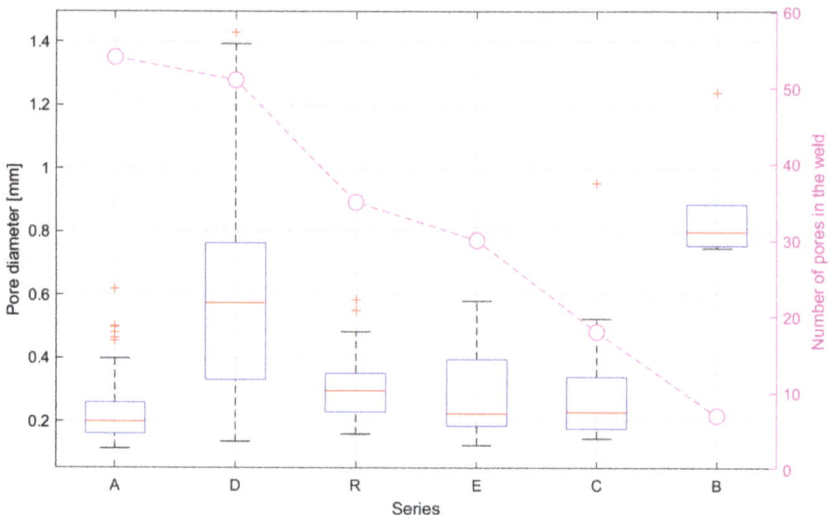

Figure 6. Number of pores and their pore diameter distribution from the second set of experiments.

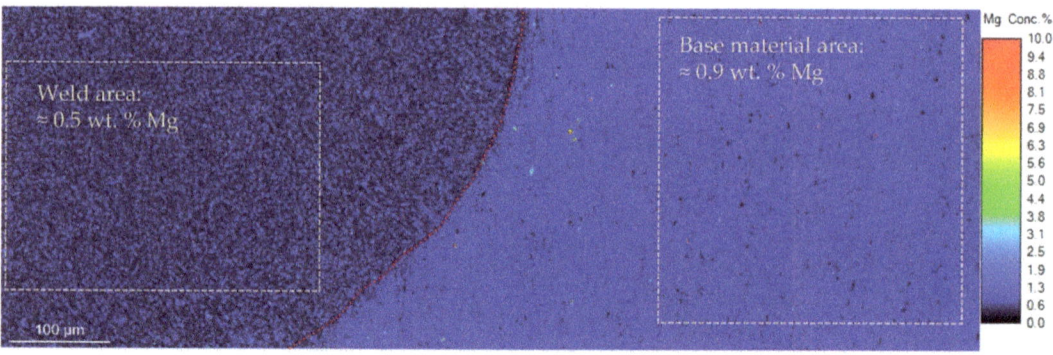

Figure 7. Detection of magnesium in the weld (**left**) via the fusion line (red dotted line) into the base material (**right**).

Very similar results were acquired around a large pore with a diameter of ≈600 μm which is shown in Figure 8. The integrated average compositions of the highlighted areas in Figures 7 and 8 revealed almost identical values for Al, Si, Mg, Fe and Mn. However, it is visible in Figure 8a,b that Al is depleted and Mg strongly enriched at the edge of the pore. The darker areas on the BSE image (indicated in Figure 8c) correspond well with the Mg-enrichment visible in Figure 8b. Further zooming into this area utilizing a SE image (Figure 8d), the Mg enriched zones display a foamy microstructure.

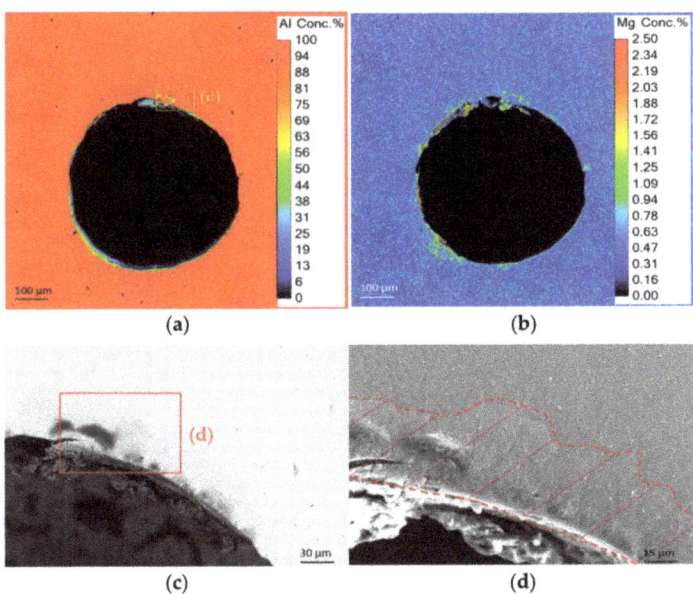

Figure 8. Elemental mapping in the area of a pore of (**a**) aluminum and (**b**) magnesium; (**c**) BSE image of pore edge; (**d**) SE image with highlighted foamy Mg-rich region.

4. Discussion

It was shown that beam oscillation has a significant effect on the pore formation in the weld of AW6082. However, it is difficult to describe the exact relationship between beam figures and pore formation. Thus, no difference in the structure of the welds was detected, which would indicate a mechanism. The structures of the welds show a wide range of equiaxed grains, which were also detected by Whang et al. [12]. They show the influence of different oscillation directions on grain growth using laser welding. Thus, the circular motion in their study showed a wide range of equiaxed grains. Since the used figures are based on superimposed oscillations with which the circle is also parameterized, a wide range of equiaxed grains was also detected in this study.

It is evident that the resulting energy fields can promote the outgassing of bubbles. Similar results regarding beam deflection and porosity were also shown by Kabasakaloglu et al. [13] and Chen et al. [14] in their investigation of laser welding, although they only considered one figure ("8" & "∞"). However, by varying different welding parameters, such as focus and frequency, they also presented different modes of energy input.

In this study, it was found that a partial shift of the maximum energy input towards the center of the weld had a positive effect on reducing the porosity, as in the case of series C. Minor improvement regarding pore size and amounts were obtained by using a two-bath technique (series E) in comparison to the reference figure. Surprisingly, series A and B, which should act as a link between a one and two bath technique, gave contradictory results in terms of pore size and number. Since A and B were performed using the same figure type, it is clear that the energy distribution significantly influences the porosity, because the figures have different functions in the energy distribution and are also different in size.

Using a narrower figure with the intention to produce a narrower weld in order to provide less space for pores to be formed did not prove to be effective. This finding is contradictory to the ones reported by Chen et al. [14]: in their study, the narrowest amplitude (0.7 mm) also had the lowest porosity (14%). However, the narrowest amplitude from Chen et al. [14] was 1.5 larger than the standard amplitude of 0.45 mm used in this study. In the aforementioned studies, welds with different parameters and beam oscillations were investigated, but this was limited to one weld per parameter set, which was never longer than 200 mm. These experiments are comparable to the first set of experiments in this study, where welds with 220 mm were also examined with regard to porosity. In the second set of experiments, it could be shown that both low and high porosity (series D) can be welded reproducibly.

Kabasakaloglu et al. [13] and Chen et al. [14] showed that frequency affects both porosity and seam geometry. Both studies showed that with higher frequency, the ratio of weld-width to -depth became less. Both concluded that there is no stable keyhole and that it is heat conduction welding. The same conclusion is drawn from this study—the weld is very wide concerning its depth, so it is also assumed to be heat conduction welding. This fact excludes the formation of pores by a keyhole. Likewise, other pore formation mechanisms, such as the penetration of hydrogen into the melt or the formation of a gas phase between aluminum oxide and liquid aluminum, as described by Fujii et al. [7], can be ruled out as the main pore formation mechanism in our study. By mechanical removal of the oxide layer and subsequent airtight packaging, re-oxidation of the aluminum was largely suppressed. A short time window between unpacking and evacuation of the welding chamber kept the progress of oxidation to a minimum. The low oxidation and the vacuum in the chamber also minimized the contributed hydrogen, which is often seen as the main reason for pore formation in other welding processes, as described by Ardika et al. [4].

Zhan et al. [5] and Zhou et al. [6] looked specifically at the outgassing of Mg in AA6061/AA5052 in their studies on EBW and fiber laser welding. The experiments in both studies were performed with rotationally symmetrical heat sources on thicker sheets (3 and 4 mm, respectively), which resulted in wider welds. Thus, in both works, the Mg content was detected concerning the weld depth and the weld center. However, the averaged decreases of Mg were similar to the results of the present work. However, the gradient shown by both Zhan et al. [5] and Zhou et al. [6] was not detected in this work.

In our study, we demonstrated that the main mechanism of pore formation in the AW6082 alloy under the given parameters (especially vacuum pressure) is the evaporation of magnesium. Zinc plays a minor role since it is present in lower concentrations (~ 0.1 wt.%) compared to magnesium (~0.9 wt.%). A direct correlation between the outgassing of these elements and the beam pattern could not yet be recognized, since at this point of the research only the reference figure R could be investigated with EPMA. Complete avoidance of porosity, therefore, seems unlikely, and further research on this topic is required.

5. Conclusions

In this study electron beam welding with newly developed welding configurations was performed on AW6082 plates. The following points comprise the most important findings:

- The beam figure and consequently the energy input influences the porosity, but the direct relationship between beam deflection and pore formation is still unclear.
- It was possible to reproducibly avoid (Series B) and create (Series D) pores.
- The pore size will also be influenced by the beam figure. However, a direct correlation between pore number and pore size could not be established.
- Outgassing of the alloying elements magnesium and zinc were identified as the main pore formation mechanism.
- Considering that magnesium and other elements will never completely outgas from the material, pore-free welding in a high vacuum is improbable.

Author Contributions: All authors contributed equally to this work. All authors have read and agreed to the published version of the manuscript.

Funding: This research was funded by the Dobeneck Technology Foundation.

Institutional Review Board Statement: Not applicable.

Informed Consent Statement: Not applicable.

Data Availability Statement: Not applicable.

Acknowledgments: The sample preparation and X-ray examinations were carried out by Peak Technology GmbH.

Conflicts of Interest: The authors declare no conflict of interest.

Appendix A

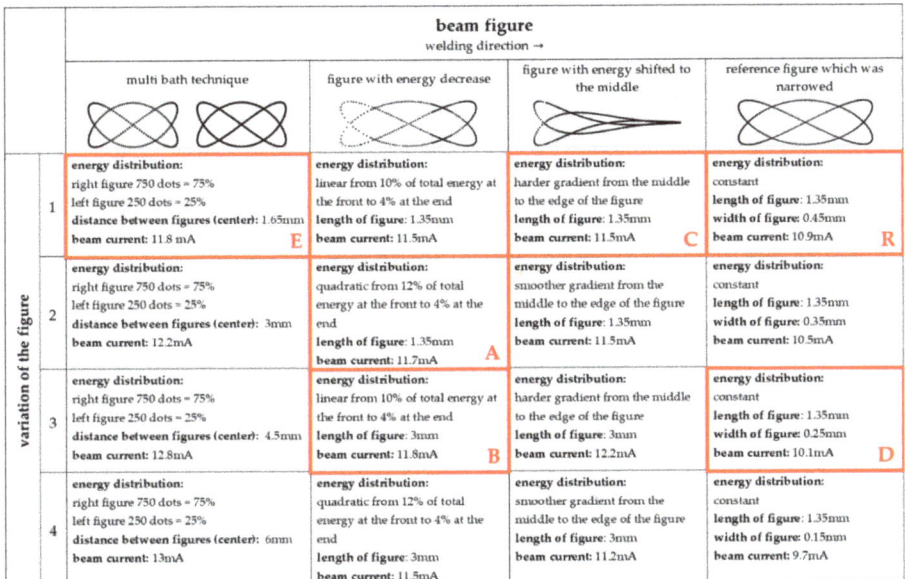

Figure A1. Test matrix from the first set of experiments. Marked in red are the six chosen figure configurations and their parameters which were applied in the second set of experiments.

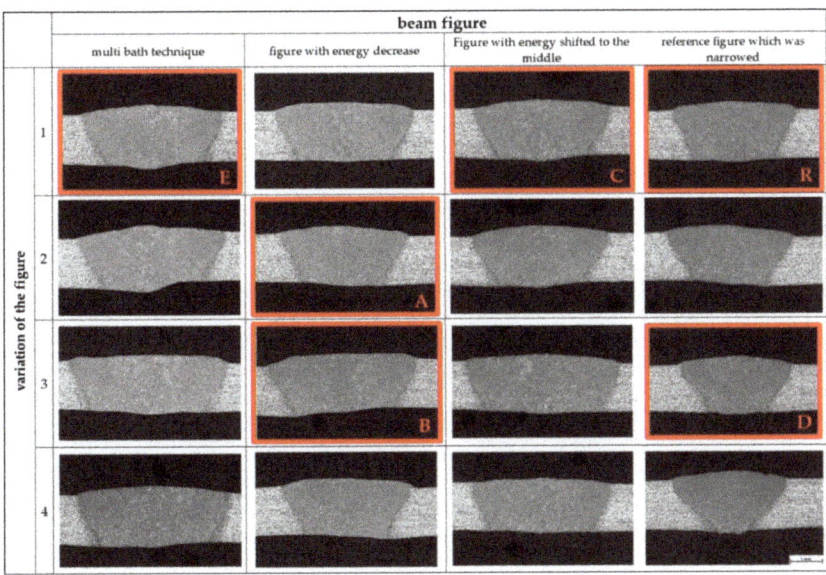

Figure A2. Cross-sections from the first set of experiments. The microstructures outlined in red are the representatives for the second set of experiments and marked like this.

Table A1. Single spot EPMA analyses of the Al-phase in the base material and in welded areas. bld.: below detection limit.

Single Spot Analysis	Al [wt.%]	Si [wt.%]	Mg [wt.%]	Fe [wt.%]	Mn [wt.%]	Cr [wt.%]	Cu [wt.%]	Zn [wt.%]	Total [wt.%]
Base material 01	96.74	0.86	0.84	0.06	0.49	0.12	0.08	0.14	99.32
Base material 02	94.47	1.53	0.78	1.35	1.19	0.19	0.12	0.14	99.77
Base material 03	97.16	0.87	0.84	0.16	0.45	0.08	0.12	0.13	99.82
Base material 04	95.56	1.16	0.81	0.67	0.92	0.20	0.07	0.15	99.55
Base material 05	97.44	0.68	0.83	bld.	0.27	0.07	0.08	0.10	99.47
Base material 06	96.47	1.01	0.85	0.38	0.56	0.10	0.10	0.11	99.58
Base material 07	97.02	0.84	0.84	0.11	0.49	0.11	0.09	0.11	99.61
Base material 08	97.23	0.76	0.83	0.05	0.44	0.16	0.07	0.15	99.69
Base material 09	94.57	1.85	1.50	0.99	0.96	0.18	0.12	0.13	100.30
Base material 10	97.02	0.82	0.81	0.09	0.49	0.12	0.10	0.10	99.55
Base material 11	96.80	0.96	0.82	0.11	0.53	0.14	0.14	0.12	99.62
Base material 12	94.62	1.43	0.80	0.90	0.81	0.11	0.10	0.16	98.93
Base material 13	96.83	0.96	0.79	0.16	0.47	0.12	0.14	0.11	99.60
Base material 14	96.99	0.86	0.81	0.13	0.44	0.10	0.09	0.14	99.56
Base material 15	95.74	1.04	0.80	0.48	0.69	0.14	0.06	0.13	99.08
Base material 16	95.06	1.31	0.79	1.16	1.01	0.17	0.11	0.15	99.76
Base material 17	96.69	0.91	0.80	0.19	0.48	0.10	0.10	0.10	99.39
Base material 18	95.31	1.16	0.80	0.64	0.80	0.16	0.08	0.13	99.08
Weld 01	96.02	1.01	0.44	0.58	0.73	0.15	0.12	bld.	99.12
Weld 02	96.08	1.28	0.56	0.50	0.71	0.13	0.12	0.09	99.47
Weld 03	96.45	1.03	0.48	0.49	0.73	0.13	0.08	bld.	99.43
Weld 04	96.29	1.07	0.47	0.43	0.66	0.13	0.11	bld.	99.22
Weld 05	96.51	1.10	0.49	0.36	0.62	0.13	0.11	bld.	99.34
Weld 06	96.28	1.24	0.45	0.44	0.67	0.17	0.08	bld.	99.39
Weld 07	96.30	1.22	0.51	0.39	0.63	0.13	0.15	0.09	99.42
Weld 08	96.43	0.95	0.45	0.55	0.71	0.13	0.11	bld.	99.38
Weld 09	97.40	0.70	0.38	0.27	0.58	0.12	0.09	bld.	99.59
Weld 10	96.36	1.04	0.43	0.57	0.75	0.16	0.08	bld.	99.46
Weld 11	95.68	1.44	0.48	0.57	0.74	0.17	0.12	bld.	99.26
Weld 12	96.02	1.56	0.50	0.54	0.71	0.14	0.09	bld.	99.57
Weld 13	96.83	1.14	0.49	0.51	0.68	0.14	0.10	bld.	99.92
Weld 14	95.46	1.53	0.46	0.69	0.76	0.14	0.11	bld.	99.19
Weld 15	96.60	1.13	0.52	0.42	0.69	0.14	0.10	bld.	99.61
Weld 16	96.42	1.34	0.56	0.47	0.64	0.16	0.07	bld.	99.71
Weld 17	95.76	1.43	0.52	0.54	0.67	0.15	0.11	bld.	99.24
Weld 18	95.67	1.57	0.52	0.51	0.68	0.14	0.09	0.07	99.24
Weld 19	95.71	1.91	0.53	0.52	0.68	0.17	0.13	bld.	99.67
Weld 20	95.69	1.10	0.48	0.52	0.69	0.11	0.07	bld.	98.73
Weld 21	96.04	1.07	0.44	0.52	0.68	0.14	0.13	0.08	99.10
Weld 22	95.34	1.60	0.49	0.45	0.68	0.16	0.10	bld.	98.83
Weld 23	96.07	1.29	0.48	0.47	0.65	0.13	0.11	bld.	99.21
Weld 24	96.01	1.30	0.54	0.44	0.67	0.13	0.09	bld.	99.25
Weld 25	96.34	0.99	0.39	0.27	0.59	0.15	0.06	bld.	98.80
Weld 26	94.98	1.95	0.49	0.57	0.72	0.14	0.11	0.08	99.03
Weld 27	95.32	1.29	0.59	0.50	0.70	0.15	0.12	0.10	98.76

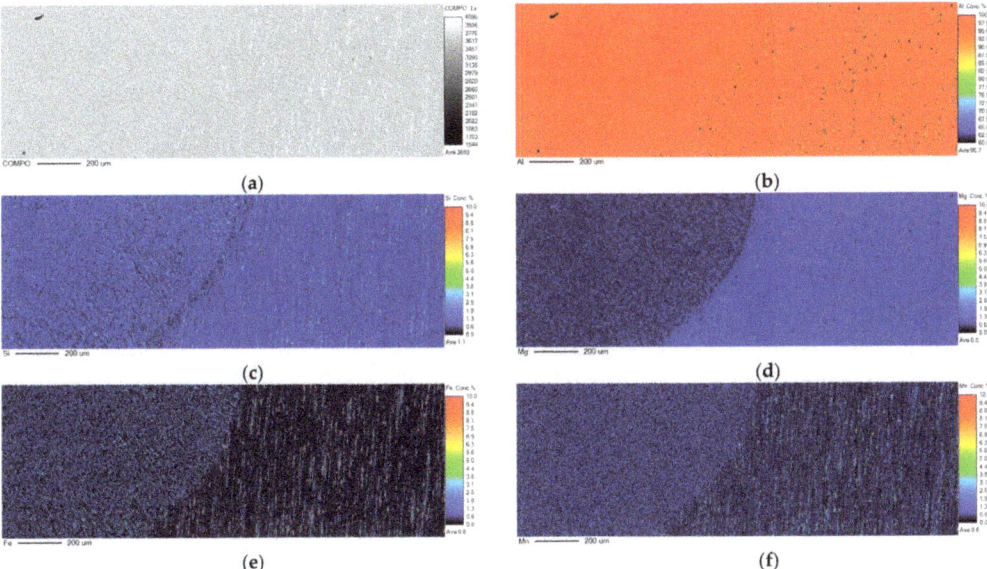

Figure A3. (**a**) BSE image and elemental distribution mappings: (**b**) Al, (**c**) Si, (**d**) Mg, (**e**) Fe, (**f**) Mn.

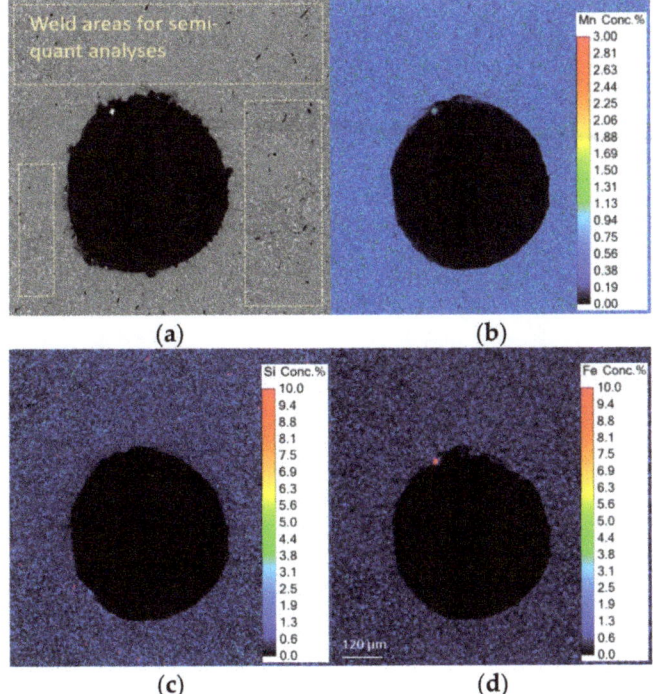

Figure A4. (**a**) BSE image showing a large pore (Ø≈0.7 mm) in a weld. The integrated area of all three indicated yellow rectangles were used to semi-quantitively evaluated the elemental mappings for Al, Mg (both shown in Figure 8), (**b**) Mn, (**c**) Si and (**d**) Fe.

References

1. Aluminium-Zentrale. *Merkblatt W7: Wärmebehandlungen von Aluminiumlegierungen*; GDA—Gesamtverband der Aluminiumindustrie e.V: Düsseldorf, Germany, 2007.
2. Barnes, T.A.; Pashby, I.R. Joining techniques for aluminium spaceframes used in automobiles Part I—Solid and liquid phase welding. *J. Mater. Process. Technol.* **2000**, *99*, 62–71. [CrossRef]
3. Rajan, R.; Kah, P.; Mvola, B.; Martikainen, J. Trends in aluminium alloy development and their joining methods. *Rev. Adv. Mater. Sci.* **2015**, *44*, 15.
4. *ECSS-Q-ST-70-39C*; ESA Requirements and Standards Division. ESA: Paris, France, 2015.
5. Ardika, R.D.; Triyono, T.; Muhayat, N. A review porosity in aluminum welding. *Procedia Struct. Integr.* **2021**, *33*, 171–180. [CrossRef]
6. Zhan, X.; Chen, J.; Liu, J.; Wei, Y.; Zhou, J.; Meng, Y. Microstructure and magnesium burning loss behavior of AA6061 electron beam welding joints. *Mater. Des.* **2016**, *99*, 449–458. [CrossRef]
7. Zhou, L.; Zhang, M.; Jin, X.; Zhang, H.; Mao, C. Study on the burning loss of magnesium in fiber laser welding of an Al-Mg alloy by optical emission spectroscopy. *Int. J. Adv. Manuf. Technol.* **2017**, *88*, 1373–1381. [CrossRef]
8. Fujii, H.; Umakoshi, H.; Aoki, Y.; Nogi, K. Bubble formation in aluminium alloy during electron beam welding. *J. Mater. Process. Technol.* **2004**, *155–156*, 1252–1255. [CrossRef]
9. Nogi, K.; Sumi, Y.; Aoki, Y.; Yamamoto, T.; Fujii, H. Welding Phenomena of Aluminum-Copper Alloy in Electron Beam Welding. *Mater. Sci. Forum* **2000**, *331–337*, 1763–1768. [CrossRef]
10. Schultz, H. *Elektronenstrahlschweißen*; DVS-Verl: Düsseldorf, Germany, 1989.
11. Wang, J.; Hu, R.; Chen, X.; Pang, S. Modeling fluid dynamics of vapor plume in transient keyhole during vacuum electron beam welding. *Vacuum* **2018**, *157*, 277–290. [CrossRef]
12. Huang, B.; Chen, X.; Pang, S.; Hu, R. A three-dimensional model of coupling dynamics of keyhole and weld pool during electron beam welding. *Int. J. Heat Mass Transf.* **2017**, *115*, 159–173. [CrossRef]
13. Wang, L.; Gao, M.; Zhang, C.; Zeng, X. Effect of beam oscillating pattern on weld characterization of laser welding of AA6061-T6 aluminum alloy. *Mater. Des.* **2016**, *108*, 707–717. [CrossRef]
14. Kabasakaloglu, T.S.; Erdogan, M. Characterisation of figure-eight shaped oscillation laser welding behaviour of 5083 aluminium alloy. *Sci. Technol. Weld. Join.* **2020**, *25*, 609–616. [CrossRef]
15. Chen, G.; Wang, B.; Mao, S.; Zhong, P.; He, J. Research on the '∞'-shaped laser scanning welding process for aluminum alloy. *Opt. Laser Technol.* **2019**, *115*, 32–41. [CrossRef]
16. Ansys Granta EduPack2020, "AL6082". Available online: https://www.ansys.com/products/materials/granta-edupack (accessed on 5 March 2022).
17. Adam, V. *Elektronenstrahlschweißen: Grundlagen Einer Faszinierenden Technik*; pro-beam: Gilching, Germany, 2011.
18. Petzow, G. *Metallographisches, Keramographisches, Plastographisches Ätzen*; 6. Auflage; Gebrüder Borntraeger: Stuttgart, Germany, 2006.
19. Bättig, D. *Angewandte Datenanalyse*; Springer: Berlin/Heidelberg, Germany, 2017. [CrossRef]

Article

Calculation of the Absorbed Electron Energy 3D Distribution by the Monte Carlo Method, Presentation of the Proximity Function by Three Parameters α, β, η and Comparison with the Experiment

Alexander A. Svintsov, Maxim A. Knyazev and Sergey I. Zaitsev *

Institute of Microelectronics Technology and High Purity Materials, Russian Academy of Sciences, Chernogolovka, St. Academician Osipyan, 6, 142432 Moscow, Russia; svintsov@iptm.ru (A.A.S.); maleksak@iptm.ru (M.A.K.)
* Correspondence: zaitsev@iptm.ru

Abstract: The paper presents a program for simulating electron scattering in layered materials *ProxyFn*. Calculations show that the absorbed energy density is three-dimensional, while the contribution of the forward-scattered electrons is better described by a power function rather than the commonly used Gaussian. It is shown that for the practical correction of the proximity effect, it is possible, nevertheless, to use the classical two-dimensional proximity function containing three parameters: α, β, η. A method for determining the parameters α, β, η from three-dimensional calculations based on MC simulation and development consideration is proposed. A good agreement of the obtained parameters and experimental data for various substrates and electron energies is shown. Thus, a method for calculating the parameters of the classical proximity function for arbitrary layered substrates based on the Monte Carlo simulation has been developed.

Keywords: electron-beam lithography; Monte Carlo method; proximity function; electrons scattering

1. Introduction

One of the most common methods for creating micro- and nanostructures in microelectronics is electron-beam lithography (EBL). Although EBL is less productive than photolithography, it turns out to be very convenient to create small structures consisting of elements of very different sizes. This makes it in demand when creating micro- and nano-objects for scientific research. Such structures can be used for studies of superconductivity [1,2], X-ray radiation [3–5], electrophysical properties of various materials (for example, graphene [6,7]) and in many other areas of physics.

One of the features of EBL is the effect of electrons back-scattered in the substrate on the electronic resist. In this case, they carry out additional exposure of the electron resist on an area usually much larger than the size of the primary electron beam. This effect is commonly called the "proximity effect" [8]. Taking into account the influence and correction of the "proximity effect" on the dose absorbed by the electronic resist allows increasing the accuracy of electron lithography, shortens the time to fabricate structures and increases the yield of a suitable product, since it reduces the sensitivity of lithography to random errors. To correct the influence of the "proximity effect" when calculating the exposure dose, the proximity function (PF) is used, i.e., the distribution of the absorbed energy in the electron resist during electron-beam scattering.

The classical PF $I(x,y)$ consists of two Gaussians [8], does not change over the depth of the resist film z, is dimensionless and is normalized to unity [9,10]. It is written in the following form:

$$I(x,y,\alpha,\beta,\eta) = \frac{\exp\left(-\left(\frac{r}{\alpha}\right)^2\right)}{\pi\alpha^2(1+\eta)} + \eta\frac{\exp\left(-\left(\frac{r}{\beta}\right)^2\right)}{\pi\beta^2(1+\eta)} \quad (1)$$

where $r^2 = x^2 + y^2$, η is the ratio of the total energy left by the reflected electrons to the energy of the forward-scattered electrons. The first Gaussian in (1) describes the distribution of the energy left in the resist by the forward-scattered electrons (characterized by parameter α), and the second one describes the distribution of the energy left in the resist by the back-scattered electrons (characterized by parameter β) [11,12].

From (1), it follows that for the practical use of PF it is necessary to know the values of the parameters α, β, η. The experience of practical correction [10,13,14] and extensive simulations [15–17] show that the three parameters found in the experiment are in most cases quite enough to obtain the required lithography accuracy.

There is a large number of works devoted to calculations [12,18–20] and to experimental measurements of PF [21–23]. The MC method [24] has long been used for determining the parameters of different functions describing the proximity effect [12,18–20] by fitting the functions to the MC calculated spatial (3D) distribution of absorbed energy. Figure 1 shows the distribution of the absorbed energy density $G(r,z)$ for a PMMA film of thickness $H_0 = 1$um on a silicon substrate, calculated by the Monte Carlo method. It can be seen that $G(r,z)$ varies in the thickness of the resist and is really a 3D function. For fitting the 2D proximity function to 3D data, usually a distribution of absorbed energy at the boundary resist–substrate (i.e., $G(r, z = H_0)$) was used [12,18–20]. However, our experience has shown that consideration of the distribution only at the resist–substrate interface or averaging over the resist thickness does not allow one to obtain PF parameters that are in good agreement with the experimental values [21]. A particularly large discrepancy arises in the determination of the η parameter. Obviously, the cause is ignoring the process of development. Experimental methods for determining the PF parameters inevitably include development processes in the measurement procedure. When considering experimental methods, we prefer methods using resist development time ("vertical" methods) [21,22], over methods using measurements of the transverse dimensions of test features such as line widths or ring widths (e.g., [23]) ("horizontal" methods). Experimental determination of PF parameters is a laborious process and can take [21] several days or even weeks. The Monte Carlo calculation takes a few minutes.

Therefore, the purpose of this work will be to develop an exclusively computational method for obtaining the parameters α, β, η of the 2D classical proximity function from the 3D absorbed energy density calculated by the Monte Carlo method with careful consideration of development. For comparison, experimental data α_e, β_e, η_e will be taken from [21].

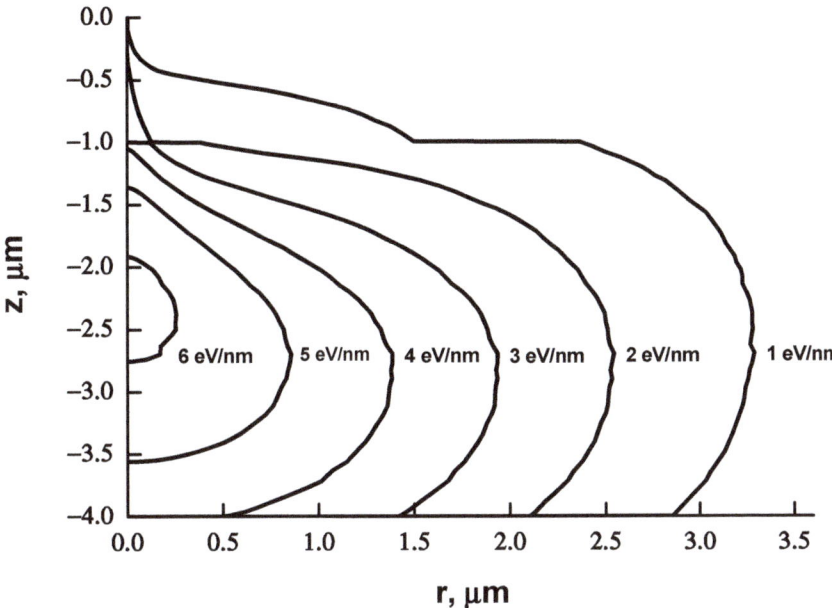

Figure 1. The results of the Monte Carlo calculation of the integral density of the absorbed energy $G_r(r, z)$. The values of the iso-levels of the density of the absorbed energy: 1, 2, 3, ... eV/nm. The resist film thickness (PMMA) is 1 micron. The substrate is Si (300 μm). The initial energy of electrons is 25 keV. The number of trajectories considered is 200,000.

2. Theory and Calculation

2.1. Calculation of Electron Scattering in Layered Materials by the Monte Carlo Method

We have developed an algorithm and implemented it in the *ProxyFn* program for the fast simulation of electron scattering in layered materials [25,26]. The structure of the algorithm is close to the one described in the work [24], but all expressions for the calculation are taken from the book by L. Reimer [27].

In the Monte Carlo simulation, it is assumed that electrons "move" in a straight line, continuously losing energy, until elastic scattering. A new direction of electron "moving" is played out using a screened Rutherford cross section [27]. The length of the straight segments is chosen randomly based on the total cross section of elastic scattering along the trajectory. Before scattering, in correspondence to the continuously slowing down approximation, the electron energy decreases, taking into account the length of the segment and the current value of stopping power (for details see [12–17,24]).

In calculations, the sample is a layered structure consisting of an arbitrary number of layers and arbitrary materials. For convenience, the sample is automatically divided into cells by a grid. In the cells of the partition grid (r_i, z_j), the energy of electrons $E(r_i, z_j)$ left by them when passing through these cells and the number of stopped electrons $N(r_i, z_j)$ are remembered. Additionally, the coefficients of reflection and transmission of electrons from the entire sample are determined, as well as the coefficients of the absorption of electrons and energy in all layers. To start the calculation, it is necessary to know only the starting energy of electrons, film thickness, chemical formulas of materials and their density. For many elements and materials, the chemical composition and density can be selected from the built-in database.

To speed up the calculation, a cylindrically symmetric and nonuniform grid r_i, z_j with center $r = 0$ on the beam axis is used. The z axis is perpendicular to the layers and directed from the outer boundary of the resist to the sample. $z = 0$, $r = 0$ is the point of entry of electrons into matter. The partition grid is set automatically.

The calculation does not consider the generation, scattering and absorption of secondary electrons, to which the fast electron gives up energy during deceleration. Although it is the secondary electrons with energies up to 50 eV that do all the "work" of breaking chemical bonds or forming bonds in resist molecules, their track length in materials does not exceed 10 nm [28] and can be taken into account by the convolution of the calculated absorbed energy density with the corresponding Gaussian. The Monte Carlo calculations also do not take into account the charging of dielectric layers, which can significantly change the trajectory of an electron. We believe that this problem can be effectively dealt with in electron lithography (for example, by applying a conductive film to the resist and grounding it, or by using a conductive resist). The PF calculation time for ten thousand trajectories takes several minutes.

2.2. Integral Proximity Function: Fitting of Absorbed Energy Distribution by Elementary Functions

It is not very difficult to implement the Monte Carlo algorithm for simulating electron trajectories. Difficulties arise when analyzing the calculation results and fitting the simulation results. After enumerating a large number of options, the following function was chosen to interpolate the density of the distribution of the electron absorbed energy $G(x,y,z)$ in the sections $z = \text{const}$:

$$G(x,y,z) = C_\delta(z)\delta(x)\delta(y) + \frac{C_a(z)}{\left(1+\left(\frac{r}{\alpha(z)}\right)^2\right)^2 \pi \alpha^2(z)} + \frac{C_b(z)\exp\left(-\left(\frac{r}{\beta(z)}\right)^2\right)}{\pi \beta^2(z)} \quad (2)$$

where r is the distance to the beam axis. The first (delta-shaped) element in expression (2) describes the electrons of the primary beam, which have not experienced one scattering on atomic nuclei. The second and third elements in (2) describe (on a qualitative level) singly and multiply scattered electrons, respectively. The coefficient $C_\delta(z)$ in the first approximation decreases exponentially with the penetration depth z, as

$$C_\delta(z) = \exp(-z/L_f)$$

where L_f is the free length. It is inversely proportional to the total cross section of electron scattering in the resist and is equal to several tens of nanometers (about 80 nm for 25 keV electrons in PMMA). In the experiment, it is not easy to separate the first and second elements in expression (2) due to the fact that the initial electron beam is not delta-shaped.

Note that the fitting parameters $C_\delta(z)$, $C_a(z)$, $C_b(z)$, $\alpha(z)$ and $\beta(z)$ depend essentially on the depth z and, in this case, the relation $\alpha(z) << \beta(z)$ is fulfilled.

To search for the fitting parameters $C_\delta(z)$, $C_a(z)$, $C_b(z)$, $\alpha(z)$ and $\beta(z)$, it is convenient to use not the distribution density $G(x,y,z)$ itself, but the integral density of the absorbed energy $G_r(r,z)$ obtained from the expression:

$$G_r(r,z) = \int_r^\infty 2\pi r' dr' G(r',z) \quad (3)$$

Note that the integral density $G_r(r,z)$ can be given a physical meaning. Consider a special structure in the form of an infinite plane with a cut out circle of radius r. It turns out that exposure of such a structure with a single dose leads to the absorbed energy density dE/dz at the center of the circle just equal to $G_r(r,z)$.

Using (2)–(3), we obtain for the three-dimensional proximity function:

$$G_r(r,z) = C_\delta(z)\Theta(r) + \frac{C_a(z)}{1+\left(\frac{r}{\alpha(z)}\right)^2} + C_b(z)\exp\left(-\left(\frac{r}{\beta(z)}\right)^2\right) \quad (4)$$

here
$$\Theta(\rho) = \begin{cases} 1, & \rho = 0 \\ 0, & \rho > 0 \end{cases}$$

On the other hand, the integrated absorbed energy density $I_r(x,y)$ in the case of the classical proximity function (1) will be equal to:

$$I_r(r,\alpha,\beta,\eta) = \int_r^\infty 2\pi r' dr' I(r',\alpha,\beta,\eta) = \frac{\exp\left(-\left(\frac{r}{\alpha}\right)^2\right) + \eta \exp\left(-\left(\frac{r}{\beta}\right)^2\right)}{1+\eta} \quad (5)$$

Figures 1 and 2 show an example of calculating the integrated absorbed energy density $G_r(r,z)$ and its approximation for a 1 µm thick PMMA e-beam resist film on a 300 µm thick silicon substrate at an initial electron energy of 25 keV. The number of trajectories considered is 200,000. The classical PF consisting of two Gaussians does not approximate $G_r(r,z)$ very well, and function (4) completes it almost ideally in all sections z.

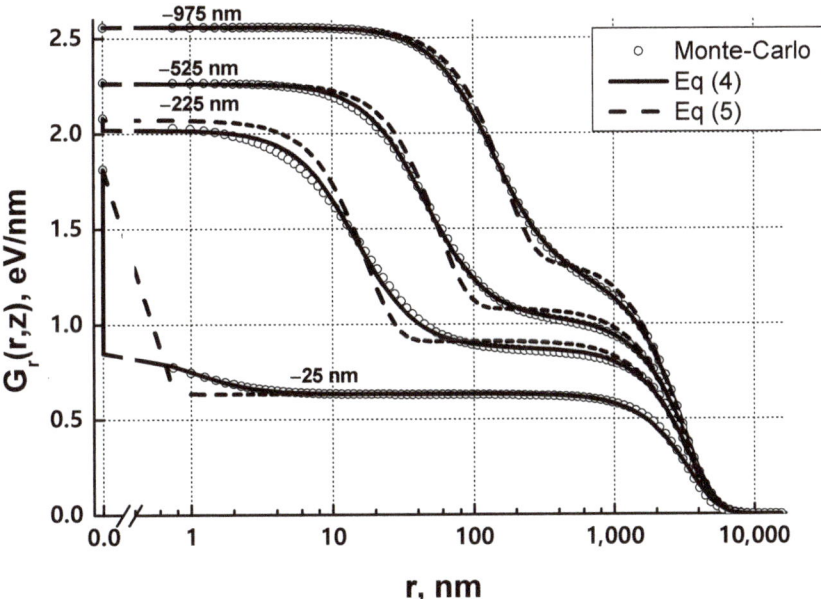

Figure 2. The results of fitting the integral density of the absorbed energy, calculated by the Monte Carlo method, by two different functions (4) and (5) for four cross sections in z: −25 nm, −225 nm, −525 nm and −975 nm. The resist film thickness (PMMA) is 1 micron. The substrate is Si (300 µm). The initial energy of electrons is 25 keV. The number of trajectories considered is 200,000.

Thus, the depth-dependent three-dimensional PF $G(r,z)$ can, in principle, be used to correct the proximity effect in e-beam lithography. However, a three-dimensional PF has significantly more fitting parameters than a classical PF with only three fitting parameters (б, в, з). This leads to a complication of calculations. On the other hand, as mentioned, the practical use (to correct the "proximity effect") of the classical PF leads to good results. Therefore, our next purpose is to determine the effective parameters of a two-dimensional PF from the three-dimensional Monte Carlo simulation results.

2.3. Fitting with Three Parameters: Analogue of Experiment

An experimental method for determining the classical PF $I(x,y,\alpha,\beta,\eta)$ (1) was proposed in [21]. The idea of the method is to search for such parameters (б, в, з) so that the test structure, consisting of elements of different sizes, after correcting the proximity effect (calculating the corrected dose based on the classical PF $I(x, y, \alpha, \beta, \eta)$) and exposure the positive resist, will be revealed exactly to the substrate in the center of each element. This experimental method and the measured parameters б, в, з from [21] have been used for more than 20 years to correct the proximity effect in the NanoMaker software and hardware complex for electron-beam lithography (www.nanomaker.com, accessed on 26 May 2022) with consistently good results.

A similar method to search for the parameters of the classical PF $I(x,y,\alpha,\beta,\eta)$ is used in this work. As in the experimental method, the calculation of the exposure dose $T(x,y)$ (electron density per unit area) is based on the classical PF $I(x,y,\alpha,\beta,\eta)$ from expression (1), but the simulation is performed instead of actual exposure and development. The absorbed dose (density of absorbed energy per unit volume) $D(x,y,z)$ in the simulation is calculated based on the three-dimensional PF $G(x,y,z)$ obtained by Monte Carlo from expression (2). For the dimensionless classical PF $I(x,y,\alpha,\beta,\eta)$, the distribution of the absorbed dose $D(x,y)$ is as follows:

$$D(x,y)/D_0 = \iint I(x-x', y-y', \alpha, \beta, \eta) T(x',y')/T_0 dx' dy' \qquad (6)$$

The development of a positive e-beam resist is simulated in the approximation of isotropic, local etching [29,30]. Then, the development rate V can be written as follows:

$$V/V_0 = (D/D_0)^\gamma \qquad (7)$$

where γ is the contrast of the resist; and D_0, V_0 are the technological constants. For a positive e-beam resist, the sensitivity T_0 is defined as the exposure dose at which an element with dimensions much larger than в is revealed in the center exactly to the substrate.

A brief description of the approach presented in Appendix A is as follows. The proposed method consists in considering a number of circular elements of different sizes R. Exposing a circle with a uniform exposure dose results in an absorbed dose distribution with a maximum exactly at the center of any circular element at all resist depths z. From isotropic local etching theory [26,30], it follows that the development front reaches the substrate for the first time namely at the center of the round. Development times T_R and T_R^i calculated for two different exposure models (for classical PF (5) and for three-dimensional PF (4)) are dependent on element radius R. Due to the simplicity of the elements, these times can be easily calculated by formulas. Further, such parameters of the classical PF б, в, з are searched with a special procedure that minimizes the objective function (9) using the ratio of the times T_R/T_R^i.

To search for б, в, з, a set of 10 round elements of radius R_n ($n = 1, \ldots, 10$) was used. The R_n value varied from the minimum value $\alpha(z)$ to the maximum value $\beta(z)$, $0 \le z \le H_0$, where $\alpha(z)$, $\beta(z)$ are the interpolation parameters of the integrated density of the absorbed energy $G_r(r,z)$.

The Appendix (A4) shows that the exposure dose ratio

$$\frac{T_R}{T_R^i} = \frac{\left(\int_0^{H_0} \frac{dz}{G_r^\gamma(0,z)}\right)^{1/\gamma}}{(1 - I_r(R, \alpha, \beta, \eta))\left(\int_0^{H_0} \frac{dz}{(G_r(0,z) - G_r(R,z))^\gamma}\right)^{1/\gamma}} \qquad (8)$$

does not contain technological parameters D_0, V_0, t_0 and can be calculated relatively easily. The exposure dose ratios (8) were calculated (for the given parameters б, в, з) for the entire set of circles R_n, and the objective function was composed from them.

$$S = \sum_n \left(\frac{T_{R,n}}{T_{R,n}^i} - 1\right)^2 \qquad (9)$$

Further, the values of the parameters α_s, β_s, η_s were determined by minimizing $S(\alpha, \beta, \eta)$. This method allows us to calculate quickly the parameters of a classical PF, depending on the thickness of the resist, the type of substrate (including the possible layered structure of the substrate), electron energy, etc.

3. Results and Discussion

A comparison of the experimental parameters α_e, β_e, η_e and those obtained from the simulation based on Monte Carlo calculations α_s, β_s, η_s 6s is necessary to verify the correctness of our chosen physical and mathematical models of electron scattering and resist development, as well as to verify the accuracy of calculations. It seems to us that a quantitative assessment of the accuracy of the Monte Carlo calculation (at least for thick resists) has not been performed before.

The experimental data were taken from [21]. In the experimental method used in this work, each of the three parameters α_e, β_e, η_e was measured in a separate test. All tests used a PMMA electronic positive resist (chemical formula $CH_2C(CH_3)(COOCH_3)$, density 1190 kg/m³). In calculations based on the Monte Carlo method, all three parameters α_s, β_s, η_s were searched simultaneously.

First, the data on the beam size 6 will be compared, and then the results of calculating в and з. The value of 6 depends on the energy, thickness and material of the resist and does not depend at all on the type of a substrate. In addition, the experimental value of 6_e is influenced by the initial beam size 6_0, which is determined by the electron microscope setting (focusing, astigmatism) and beam jitter. In fact, the value of 6_s obtained by the Monte Carlo method should be compared with parameter $\sqrt{\alpha_e^2 - \alpha_0^2}$.

Table 1 shows a comparison of the size of the forward-scattered electron beam 6, obtained from the 6_e experiment (except the initial beam size 6_0) and calculated on the basis of the Monte Carlo method 6_s for three energies E = 15, 25 and 35 keV and a set of resist thicknesses H_0 = 100, 200, 500, 1000 and 1500 nm on a silicon substrate. The experimental data 6_e were interpolated by the formula $\alpha_e^2 = A_E H_0^3 / E^2 + \alpha_0^2$, where the constants A_E and the initial beam dimensions 6_0 for different energies E were obtained from the experiment [21]. In the Monte Carlo calculation, the beam was assumed to be absolutely thin. Table 1 shows that the experimental data for 6_e and the results of calculating 6_s are in good agreement.

Table 1. Comparison of the parameters of the proximity function αs, (obtained by the method described above based on the Monte Carlo calculation) and αe, (calculated from the results of interpolation of experimental data [21]), for three values of the electron energy E and different thicknesses H_0 of the PMMA resist. The substrate is Si.

E, keV	15		25		35	
H_0, nm	α_e, nm	α_s, nm	α_e, nm	α_s, nm	α_e, nm	α_s, nm
100	6	4	3	2	2	1
200	16	12	10	6	7	5
500	64	56	39	30	28	20
1000	182	183	110	96	79	65
1500	334	342	202	196	145	130

A comparison of the calculated (β_s, η_s) and experimental (β_e, η_e) for various substrates and accelerating voltages can be carried out using the data in Tables 2–4. For all cases, a positive PMMA resist 500 nm thick with a contrast γ = 3 was used. The values of α_s, β_s, η_s turned out to be stable with respect to the change in contrast, and hardly changed for г = 2.5, 3 or 4. For comparison, the experimental data for Si, GaAs, Al$_2$O$_3$ and mica from [21] were used; the data for Ge and C (diamond) substrates were specially measured for this work by the method [21]. The calculated values of β_s with an accuracy of ŷ10% coincided with the experimental values of β_e, for з the accuracy was ŷ25%.

Table 2. Comparison of the parameters of the proximity function obtained in the experiment β_e, η_e and as a result of calculating β_s, η_s for different energies of electrons E and Si and GaAs substrates. The PMMA resist (ERP-40) 0.5 μm thick.

Substrate	Si				GaAs			
Density	2330 kg/m³				5350 kg/m³			
E, keV	β_e	β_s	η_e	η_s	β_e	β_s	η_e	η_s
11	0.9	0.85	-	0.93	-	0.73	-	1.23
15	1.5	1.33	-	0.87	-	0.92	-	1.24
20	2.2	2.11	-	0.79	1.2	1.17	-	1.23
25	3.1	3.01	0.7	0.73	1.5	1.48	1.4	1.16
30	4	4.08	-	0.69	2	1.85	-	1.11
35	5.8	5.32	-	0.66	2.3	2.29	-	1.07
39	-	6.39	-	0.63	2.6	2.66	-	1.04

Table 3. Comparison of the parameters of the proximity function obtained in the experiment β_e, η_e and as a result of calculating β_s, η_s for different energies of electrons E, Al$_2$O$_3$ substrates and mica. The PMMA resist (ERP-40) 0.5 μm thick.

Substrate	Al$_2$O$_3$				KAl$_2$Si$_3$O$_{10}$(OH)$_2$ (mica)			
Density	3970 kg/m³				2850 kg/m³			
E, keV	β_e	β_s	η_e	η_s	β_e	β_s	η_e	η_s
11	-	0.76	-	0.76	0.75	0.81	-	0.81
15	1.	1.02	-	0.72	1.2	1.19	-	0.75
20	-	1.47	-	0.65	2	1.82	-	0.62
25	2	2.05	0.8	0.59	2.7	2.58	0.5	0.61
30	-	2.71	-	0.56	3.7	3.53	-	0.59
35	3.4	3.48	-	0.53	4.8	4.59	-	0.56
39	-	4.19	-	0.52	-	5.54	-	0.54

Table 4. Comparison of the parameters of the proximity function obtained in the experiment β_e, η_e and as a result of calculating β_s, η_s for different energies of electrons E, Ge substrates and diamond. The PMMA resist (ERP-40) 0.5 μm thick.

Substrate	Ge				C (Diamond)			
Density	5323 kg/m³				3500 kg/m³			
E, keV	β_e	β_s	η_e	η_s	β_e	β_s	η_e	η_s
11	-	0.73	-	1.26	0.7	0.79	-	0.51
15	0.7	0.92	-	1.28	1.0	1.08	-	0.41
20	1.1	1.15	-	1.24	1.6	1.61	-	0.33
25	1.4	1.46	1.1	1.18	2.1	2.23	0.4	0.3
30	1.8	1.84	-	1.12	2.6	2.96	-	0.29
35	2.5	2.27	-	1.08	3.6	3.87	-	0.26
39	-	2.67	-	1.06	-	4.63	-	0.25

Note that the parameters в and з did not depend on focusing (if $α_s \ll β_s$) and were determined only by the properties of the resist, substrate and the initial energy of electrons; therefore, they have a fundamental value.

4. Conclusions

The algorithm for the fast calculation of the absorbed energy density of electrons $G(r)$ by the Monte Carlo method for layered materials was described.

To interpolate the calculated absorbed energy density of electrons $G(r)$, a fitting function (2) was proposed, which described well the distribution of the absorbed energy of electrons in layered materials depending on the distance to the center of the beam r and on the depth z. The power member describing the scattering of primary electrons seemed to be nontrivial and was considered for the first time.

A numerical procedure was proposed that takes into account the development of the resist and makes it possible to replace the complex 3D distribution of the absorbed energy with a classical (two-dimensional) proximity function with three parameters б, в, з.

The examples of calculating the parameters $α_s, β_s, η_s$ of the proximity function were shown for different energies of electrons and substrates and their comparison with the experimental data $α_e, β_e, η_e$. Calculations of в with an accuracy of ŷ10% coincided with the experiment; for з, the accuracy was ŷ25%.

Thus, it can be argued that the experimental confirmation of the accuracy of calculating PF by the Monte Carlo method and the procedure of interpolating PF with three parameters (б, в, з) has been obtained.

Author Contributions: Conceptualization, A.A.S. and S.I.Z.; software, A.A.S.; validation, A.A.S., M.A.K. and S.I.Z.; investigation, M.A.K.; data curation, M.A.K.; writing—original draft preparation, A.A.S., writing—review and editing, M.A.K., S.I.Z.; supervision, A.A.S., S.I.Z. All authors have read and agreed to the published version of the manuscript.

Funding: This research was funded by the Ministry of Sciences and Higher Education of the Russian Federation, grant number (075-00355-21-00). The APC was funded by IMT RAS.

Institutional Review Board Statement: Not applicable.

Informed Consent Statement: Not applicable.

Data Availability Statement: Data underlying the results presented in this paper are not publicly available at this time but may be obtained from the authors upon reasonable request.

Conflicts of Interest: The authors declare no conflict of interest.

Appendix A

The ratio of exposure doses is found for classical and three-dimensional proximity functions. The density of absorbed energy (dose) in the center of a circle with the radius R, exposed with a constant dose T (density of electrons per unit area), is expressed through the integral proximity function:

$$D(R,z) = (G_r(0,z) - G_r(R,z))\, T \qquad (A1)$$

For the classical dimensionless PF $I(x,y,б,в,з)$, the absorbed energy distribution is written as follows:

$$\frac{D}{D_0} = (I_r(0,α,β,η) - I_r(R,α,β,η))\frac{T}{T_r} = (1 - I_r(R,α,β,η))\frac{T}{T_0} \qquad (A2)$$

In order for the resist to reveal exactly in the center of the circle to the substrate, the following condition should be held: $D/D_0 = 1$.

Then, it follows from (A2) that the exposure dose T_R should be equal to:

$$T_R = \frac{T_0}{1 - I_r(R, \alpha, \beta, \eta)}$$

The "ideal" exposure dose T_R^i is calculated using the three-dimensional PF. In the center of each element, the absorbed dose has a maximum on the XY plane in each section $z = \text{const}$; therefore, a positive electron resist is revealed vertically along the z axis in the center of the circle [30]. The absorbed dose $D_R(z)$ in the center of a circle with the radius R_n exposed with a certain dose T is obtained from expression (A1). The time t for the development of a positive resist with a thickness H_0 to the substrate in the center of the circle is $t = \int_0^{H_0} \frac{dz}{V(z)}$. Then for a circle with the radius R:

$$t_R = \left(\frac{D_0}{T}\right)^\gamma \int_0^H \frac{dz}{(G_r(0,z) - G_r(R,z))^\gamma V_0} \quad (A3)$$

The "ideal" exposure dose T_R^i can be obtained from expression (A3), at which a circle with the radius R in the center is revealed to the bottom for a given development time t_0:

$$T_R^i = \frac{D_0}{(V_0 t_0)^{\frac{1}{\gamma}}} \left(\int_0^{H_0} \frac{dz}{(G_r(0,z) - G_r(R,z))^\gamma}\right)^{\frac{1}{\gamma}}$$

The sensitivity of the positive resist T_0 can also be calculated using this expression at $R = \infty$:

$$T_0 = \frac{D_0}{(V_0 t_0)^{\frac{1}{\gamma}}} \left(\int_0^{H_0} \frac{dz}{G_r^\gamma(0,z)}\right)^{\frac{1}{\gamma}}$$

As a result, the ratio of exposure doses T_R/T_R^i is obtained, which does not contain unknown technological parameters D_0, V_0 and t_0:

$$\frac{T_R}{T_R^i} = \frac{\left(\int_0^{H_0} \frac{dz}{G_r^\gamma(0,z)}\right)^{1/\gamma}}{(1 - I_r(R, \alpha, \beta, \eta)) \left(\int_0^{H_0} \frac{dz}{(G_r(0,z) - G_r(R,z))^\gamma}\right)^{1/\gamma}} \quad (A4)$$

References

1. Gurtovoi, V.L.; Burlakov, A.A.; Nikulov, A.V.; Tulin, V.A.; Firsov, A.A.; Antonov, V.N.; Davis, R.; Pelling, S. Multiple current states of two phase-coupled superconducting rings. *JETP* **2011**, *113*, 678–682. [CrossRef]
2. Kuznetsov, V.I.; Firsov, A.A. Unusual quantum magnetic-resistive oscillations in a superconducting structure of two circular asymmetric loops in series. *Phys. C* **2013**, *492*, 11–17. [CrossRef]
3. Knyazev, M.A.; Svintsov, A.A.; Fahrtdinov, R.R. Development of Field Alignment Methods for Electron-Beam Lithography in the Case of X-ray Bragg–Fresnel Lenses. *J. Surf. Investig. X-ray Synchrotron Neutron Tech.* **2018**, *12*, 957–960. [CrossRef]
4. Firsov, A.A.; Svintsov, A.A.; Zaitsev, S.I.; Erko, A.; Aristov, V.V. The first synthetic X-ray hologram: Results. *Opt. Commun.* **2002**, *202*, 55–59. [CrossRef]
5. Grigoriev, M.; Fakhrtdinov, R.; Irzhak, D.; Firsov, A.; Firsov, A.; Svintsov, A.; Erko, A.; Roshchupkin, D. Two-dimensional X-ray focusing by off-axis grazing incidence phase Fresnel zone plate on the laboratory X-ray source. *Opt. Commun.* **2017**, *385*, 15–19. [CrossRef]
6. Novoselov, K.S.; Geim, A.K.; Morozov, S.V.; Jiang, D.; Katsnelson, M.I.; Grigorieva, I.V.; Dubonos, S.V.; Firsov, A.A. Two-dimensional gas of massless Dirac fermions in graphene. *Nature* **2005**, *438*, 197–200. [CrossRef]
7. Morozov, S.V.; Novoselov, K.S.; Schedin, F.; Jiang, D.; Firsov, A.A.; Geim, A.K. Two-Dimensional Electron and Hole Gases at the Surface of Graphite. *Phys. Rev. B* **2005**, *72*, 201401. [CrossRef]
8. Chang, T.H.P. Proximity effect in electron-beam lithography. *J. Vac. Sci. Technol.* **1975**, *12*, 1271–1275. [CrossRef]

9. Parikh, M. Corrections to proximity effects in electron–beam lithography. 1. Theory. *J. Appl. Phys.* **1979**, *50*, 4371–4377. [CrossRef]
10. Aristov, V.V.; Gaifullin, B.N.; Svintsov, A.A.; Zaitsev, S.I.; Jede, R.R.; Raith, H.F. Accuracy of proximity correction in electron lithography after development. *J. Vac. Sci. Technol.* **1992**, *10*, 2459–2467. [CrossRef]
11. Greeneich, J.S.; Van Duzer, T. An exposure model for electron-sensitive resists. *IEEE Trans. Electron.* **1974**, *21*, 286–299. [CrossRef]
12. Parikh, M.; Kyser, D. Energy deposition functions in electron resist films on substrates. *J. Appl. Phys.* **1979**, *50*, 1104–1111. [CrossRef]
13. Available online: https://www.raith.com/technology/nanofabrication-software/proximity-effect-correction/ (accessed on 26 May 2022).
14. Available online: https://www.genisys-gmbh.com/part-1-electron-scattering-and-proximity-effect.html (accessed on 26 May 2022).
15. Lee, S.-Y.; Dai, Q.; Lee, S.-H.; Kim, B.-G.; Cho, H.-K. Enhancement of spatial resolution in generating point spread functions by Monte Carlo simulation in electron-beam lithography. *J. Vac. Sci. Technol. B* **2011**, *29*, 06F902. [CrossRef]
16. Li, P. A Review of Proximity Effect Correction in Electron-beam Lithography. *arXiv* **2015**, arXiv:1509.05169.
17. Figueiro, T.R. Process Modeling for Proximity Effect Correction in Electron Beam Lithography. Micro and Nanotechnologies/Microelectronics. Doctoral Dissertation, Université Grenoble Alpes, Gières, France, 2015.
18. Fretwell, T.A.; Gurung, R.; Jones, P.L. Curve fittin to Monte Carlo data for the determination of proximity effect correction parameters. *Microelectron. Eng.* **1992**, *17*, 389–394. [CrossRef]
19. Katia, V.; Georgi, M. Modelling of exposure and development processes in electron and ion lithography. *Model. Simul. Mater. Sci. Eng.* **1994**, *2*, 239–254. [CrossRef]
20. Katia, V.; Georgi, M. Computer Simulation of Processes at Electron and Ion Beam Lithography. In *Lithography*; Wang, M., Ed.; Intech: Rijeka, Croatia, 2010; pp. 319–350. [CrossRef]
21. Aparshina, L.I.; Dubonos, S.V.; Maksimov, S.V.; Svintsov, A.A.; Zaitsev, S.I. Energy dependence of proximity parameters investigated by fitting before measurement tests. *J. Vac. Sci. Technol. B* **1997**, *15*, 2298–2302. [CrossRef]
22. Rooks, M.; Belic, N.; Kratschmer, E.; Viswanathan, R. Experimental optimization of the electron-beam proximity effect forward scattering parameter. *J. Vac. Sci. Technol. B* **2005**, *23*, 2769–2775. [CrossRef]
23. Stevens, L.; Jonckheere, R.; Froyen, E.; Decoutere, S.; Lanneer, D. Determination of the proximity parameters in electron beam lithography using doughnut-structures. *Microelectron. Eng.* **1986**, *5*, 141–150. [CrossRef]
24. Hawryluk, R.; Hawryluk, A.; Smith, H. Energy dissipation in a thin polymer film by electron beam scattering. *J. Appl. Phys.* **1974**, *45*, 2551–2566. [CrossRef]
25. Pavlov, V.N.; Panchenko, V.Y.; Polikarpov, M.A.; Svintsov, A.A.; Yakimov, E.B. Simulation of the current induced by ^{63}Ni beta radiation. *J. Surf. Investig. X-ray Synchrotron Neutron Tech.* **2013**, *7*, 852–857. [CrossRef]
26. Zaitsev, S.I.; Pavlov, V.N.; Panchenko, V.Y.; Polikarpov, M.A.; Svintsov, A.A.; Yakimov, E.B. Comparison of the efficiency of ^{63}Ni beta-radiation detectors made from silicon and wide-gap semiconductors. *J. Synch. Investig.* **2014**, *8*, 843–845. [CrossRef]
27. Reimer, L. *Scanning Electron Microscopy. Physics of Image Formation and Microanalysis*; Springer Series in Optical Sciences; Peter, W., Ed.; Springer: Hawkes, ON, Canada, 1998; Volume 45.
28. Bronstein, I.M.; Fraiman, B.S. *Secondary Electron Emission*; Nauka: Moscow, Russia, 1969; p. 170.
29. Zaistev, S.I.; Svintsov, A.A. Theory of isotropic local etching. Problem statement and basic equation. *Poverhnost* **1986**, *4*, 27–35.
30. Zaistev, S.I.; Svintsov, A.A. Theory of isotropic local etching. Properties and some analytical solution. *Poverhnost* **1987**, *1*, 47–56.

Article
Recycling of Technogenic CoCrMo Alloy by Electron Beam Melting

Katia Vutova [1,*], Vladislava Stefanova [2], Vania Vassileva [1] and Stela Atanasova-Vladimirova [3]

1. Institute of Electronics, Bulgarian Academy of Sciences, 1784 Sofia, Bulgaria; vvvania@abv.bg
2. Department of Metallurgy of Non-Ferrous Metals and Semiconductors Technologies, University of Chemical Technology and Metallurgy, 1756 Sofia, Bulgaria; vps@uctm.edu
3. Academician Rostislav Kaishev Institute of Physical Chemistry, Bulgarian Academy of Sciences, 1113 Sofia, Bulgaria; statanasova@ipc.bas.bg
* Correspondence: katia@van-computers.com

Abstract: In the current work, the possibility of the recycling of technogenic CoCrMo material by electron beam melting is investigated. The influence of thermodynamic and kinetic parameters (temperature and melting time) on the behavior of the main components of the alloy (Co, Cr, and Mo) and other elements (Fe, Mn, Si, W, and Nb) present in it, and on the microstructure of the ingots obtained after e-beam processing is studied. The vapor pressure of the alloy is determined taking into account the activities of the main alloy components (Co, Cr, and Mo). The relative volatility of the metal elements present in the alloy was also evaluated. An assessment of the influence of the temperature and the retention time on the degree of elements removal from CoCrMo technogenic material was made. The results obtained show that the highest degree of refining is achieved at 1860 K and a residence time of 20 min. The conducted EDS analysis of the more characteristic phases observed on the SEM images of the samples shows distinct micro-segregation in the matrix composition.

Keywords: technogenic Co–Cr–Mo alloy; electron beam recycling; refining process; degree of removal

1. Introduction

Co-based alloys have been used for more than 80 years as metal biomaterials in dental and surgical prosthetics. In recent years, Co–Cr alloys, additionally alloyed with Mo, W, Ni, and Ti, have become widely used due to their high biological tolerance, low carcinogenicity, excellent mechanical properties and high corrosion resistance [1–6]. An improvement in the composition of the alloy can also be achieved by adding some alloying elements such as Si, Nb, and Ir in low concentrations, heat treatments, and developing new processes for production and casting of alloys [3].

Co-based alloys can be classified into two main groups cast—Co–Cr–Mo alloys and hot forged Ni–Co–Cr–Mo alloys. They usually contain 58–69% Co, 26–30% Cr, 5–7% Mo, and other metals such as Ni, Fe, Mn, Si, and W in precisely defined amounts and they are also known as ASTM F75 alloys [7]. Conventional Ni–Co–Cr–Mo alloys are widely used in prosthetics and in the manufacture of the so-called "stents" used for coronary heart disease. They are additionally alloyed with C for greater wear resistance. The main problem with these alloys is still their low ductility and the release of nickel ions during operation.

Cobalt is a metal that determines the basic mechanical properties of the alloy, such as hardness, strength, and toughness [8]. Chromium, whose content is up to 30%, provides biocompatibility and corrosion resistance by forming a protective oxide layer (Cr_2O_3) that prevents the diffusion of metal atoms and their contact with oxygen. The molybdenum content in Co–Cr–Mo alloys is usually about 5%. Molybdenum protects the alloy from the action of halogens and their compounds. Together with manganese, it contributes to the fluidity of the alloy. Iron is an inevitable component of the alloy. It improves the machining of the alloy and reduces its electrochemical stability. Silicon, like manganese, contributes to the detoxification and cleansing of the alloy during melting. Carbon protects the alloy from oxidation. However, its content must be strictly controlled (<0.35%) because it tends to form carbides with the metals present in the alloy, thus negatively affecting the basic properties of the alloy such as elasticity and hardness.

There are a number of studies in the literature on the microstructure of Co–Cr–Mo alloys, taking into account the differences in the composition of the samples, the type, and content of the alloying elements, the methods of melting and hardening, the melting temperature, the retention time, etc. [3,6–14].

Co–Cr–Mo alloys can be obtained both by conventional casting processing and by other alternative methods such as selective laser and electron beam melting, computer-aided design/computer-aided manufacturing (CAD/CAM) and others [10,11,13–17]. The electron beam melting (EBM) method, which combines the advantages of vacuum and high-energy special electrometallurgy, deserves special attention [18–20]. The method allows for the removal of components and ensures the production of pure metals and alloys in the processing of technogenic materials [21–24].

In this work, the possibility of recycling of technogenic CoCrMo material (waste from the dental technology) by electron beam melting to find technological solutions for refining the alloy is investigated. For this purpose, the influence of thermodynamic and kinetic parameters (treatment temperature and refining time) on the behavior of the main components and the alloying elements (W, Fe, Si, Mn, and Nb), and on the microstructure of the ingots obtained after electron beam melting is studied. Based on thermodynamic analysis and conducted experiments, an assessment of the efficiency of the refining process during EBM was made.

2. Material and Methods

The experiments for CoCrMo technogenic material melting were conducted using EBM furnace ELIT-60 (Leybold GmbH, Cologne, Germany) with power 60 kW at the Physical problems of the e-beam technologies laboratory of the Institute of electronics, Bulgarian Academy of Sciences. ELIT-60 is equipped with a melting chamber and one electron gun with an accelerating voltage of 24 kV. The refined metal material solidifies in a water-cooled copper crucible with moving bottom [18,22] and the operation vacuum pressure is 1×10^{-3} Pa.

The tested material is a technogenic CoCrMo alloy—waste from the dental technology, used in the dental practice for the manufacture of dentures, implants and other products (Figure 1).

Figure 1. Technogenic CoCrMo material.

Data about the chemical composition of the investigated CoCrMo alloy before EBM is presented in Table 1. The chemical composition of Co–Cr alloy (F75-12) registered in the ASTM standards for biomedical applications [1] is also shown in Table 1.

The table shows that the investigated technogenic material differs from the ASTM F75-12 standard in the chemical composition of the main components and the reduced content of Co and Mo and the high content of the Cr, W, and Si elements require its refining.

A refining of the source material was performed at electron beam power (P_b) of 2.25 kW (T = 1790 K), 3.75 kW (T = 1830 K), 4.50 kW (T = 1845 K), 4.75 kW (T = 1855 K), and 5.0 kW (T = 1860 K) and refining duration of 10 min, 20 min, and 30 min. For each technological mode the changes in the composition and structure of the material after the e-beam process were controlled. The degree of refining of each of the controlled elements for each technological mode was calculated.

Table 1. Chemical composition (mass%) of the investigated technogenic CoCrMo material and Co-Cr alloy (F75-12) according to the ASTM standards.

Sample	Co	Cr	Mo	Ni	Fe	C	Mn	Nb	W	Si	Others
Initial material	61.0	31.22	4.78	0.0	0.65	0.0	0.43	0.32	0.38	1.09	0.13
ASTM F75-12	balance	27–30	5–7	<0.5	<0.75	<0.35	<1.0	-	<0.2	<1.0	<0.49 [1]

[1] $P < 0.02$, $S < 0.01$, $N < 0.25$, $Al < 0.1$, $Ti < 0.1$, $B < 0.01$.

The temperature was determined by an optical pyrometer QP-31 using special correction filters.

The chemical composition of the source material and the specimens after the EBM was determined by emission spectral analysis. The baseline and final concentrations were controlled for both the main components Co, Cr, and Mo as well as for the Fe, Mn, Nb, W, and Si elements.

The preparation of the samples for the metallographic study includes standard procedure, grinding, polishing, and etching. A reagent Glyceregia prepared from 15 mL of HCl, 10 mL glycerol, and 5 mL HNO_3 [25] was used for etching. The time to manifest the macro and microstructure of the examined samples is ~20 min.

Light microscopy and scanning electron microscopy (SEM) are employed to investigate the macro and microstructure of the ingots (in the centre and periphery) obtained after the EBM.

A light microscope Leica DM2500 (Leica Microsystems GmbH, Wetzlar, Germany) with a digital camera Leica EC3 (Leica Microsystems GmbH, Wetzlar, Germany) was used for the topographic study of the macrostructure after e-beam processing of the specimens. The Leica LAS software (Leica Microsystems GmbH, Wetzlar, Germany) was used for image processing.

The microstructure of CoCrMo alloys was investigated using a Scanning Electron Microscope JEOL 6390 with INCA Oxford EDS detector and an elemental chemical analysis of the composition of the phases observed on the SEM/BEC images was made by Energy Dispersive Spectroscope (EDS) analysis.

3. Results and Discussion

3.1. Thermodynamic Conditions of Element Volatilization during Electron Beam Melting and Refining

Depending on the thermodynamic conditions of the EBM process and on the type of the removed component, the refining processes could be realized through the following methods: degassing—removal of components with a partial pressure, which is higher than the vapor pressure of the base metal; and distillation—evaporation of more volatile compounds of the metallic components [18].

Figure 2 shows the vapor pressure values of the pure metals (Co, Cr, Mo, Mn, Fe, Si, Nb, and W) present in the studied alloy. They are calculated for a temperature range from 1700 K to 2000 K and an operating pressure in the vacuum chamber of 1×10^{-3} Pa. The calculations were made using the professional thermochemical calculation programme HSC Chemistry ver.7.1, module "Reaction Equation" [26].

The figure shows that elements such as W and Nb have a significantly lower vapor pressure than that of the main components of the alloy (Co, Cr, and Mo). Therefore, these elements cannot be removed from the alloy during the EBM. Out of the other controlled elements (Si, Fe, and Mn) only manganese has a vapor pressure higher than that of Co, Cr, and Mo. The iron has a vapor pressure higher than that of Co and Mo and close to that of Cr, while the vapor pressure of silicon is close to that of cobalt. Therefore, in the tested temperature range, these elements can be removed from the reaction surface liquid material/vacuum. Out of the main components of the alloy, chrome has the highest vapor pressure while that of Mo is very low. Therefore, more intense evaporation can be expected only for two (Cr and Co) out of the three elements.

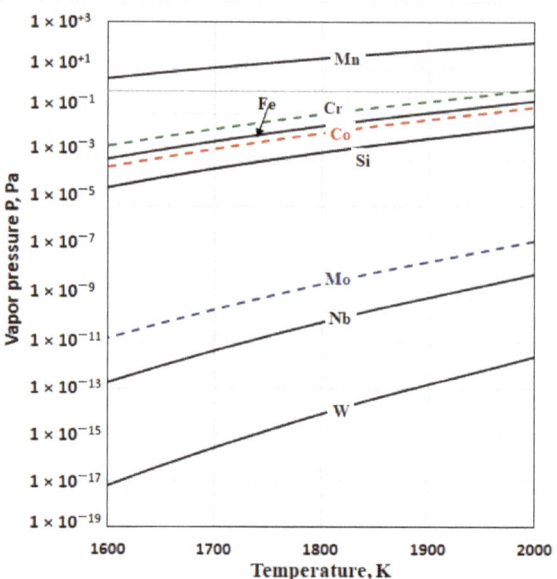

Figure 2. The vapor pressure of the pure elements as a function of the temperature in vacuum.

The criterion for assessing the effectiveness of the refining process for multi-component metal systems where the metal or metal alloy is melted in a vacuum is the relative volatility (α) [27]. It can be calculated by an equation:

$$\alpha_i = \frac{p_{alloy}}{p_i} \frac{\sqrt{M_i}}{\sqrt{M_{alloy}}} \quad (1)$$

where p_{alloy} is the cumulative vapor pressure of the main components of the alloy (Co, Cr, and Mo); and p_i—is the vapor pressure of the other elements present in the alloy (Si, Fe, Mn, W, and Nb). The molecular mass of CoCrMo alloy (M_{alloy}) is calculated from the expression:

$$M_{alloy} = M_{Co}x_{Co} + M_{Cr}x_{Cr} + M_{Mo}x_{Mo} \quad (2)$$

where M_{Co}, M_{Cr}, and M_{Mo} are the molecular masses of cobalt, chrome, and molybdenum and x_{Co}, x_{Cr}, and x_{Mo} their mols. M_i are the molecular masses of the other elements present in the alloy.

When calculating the vapor pressure (p_{alloy}) of CoCrMo alloy, it is necessary to take into account the interaction between cobalt, chrome, and molybdenum in the alloy, i.e., their activity. The actual vapor pressure ($p_{j(Me)}$) of each of these components is:

$$p_{j(Me)} = P^o_{Me} a_{j(Me)} \quad (3)$$

where P^o_{Me} is the vapor pressure of the pure metals Co, Cr, and Mo, $a_{j(Me)}$—the activities of the same components in the alloy.

A number of theoretical and experimental models for determining the activity of components in binary systems are known in the literature [28–30]. To determine the Co,

Cr and Mo activity in the alloy this study uses the methodology based on two-component diagrams and is presented in [28]. The activity of components in the alloy is calculated using the equation:

$$log(a^T_{j(Me)}) = -\frac{(T_o - T_{liq})H_f}{4574 \cdot T \cdot T_o} + \frac{T - T_{liq}}{T} log x_j \quad (4)$$

where $a^T_{j(Me)}$ is the activity of an alloy element (j) at operating temperature T; T_o—the melting temperature of an element; T_{liq}—the temperature of the liquidus surface at a molar portion (x_j) of the component in the alloy; and H_f—enthalpy of fusion of a given component.

When using Equation (4) or other similar equations, one should keep in mind that these equations are associated with parameters, which are difficult to determine and always entail an error [28]. Nevertheless, the activities calculated using these equations give approximate information about the activity of the components in a more complex metal system.

Table 2 shows the baseline data and the activities of Co, Cr, and Mo in the investigated alloy calculated according to Equation (4). The liquidus surface temperatures were determined based on the binary systems Co–Cr and Co–Mo. The calculations were performed at an operating temperature of 1860 K and a molar composition corresponding to the starting alloy.

Table 2. Starting data and activities of Co, Cr, and Mo for initial CoCrMo alloy at T = 1860 K.

Element	x_j	T_0, K	H_f, kJ/mol	T_{liq}, K	T, K	Activity, $a^T_{j(Me)}$
Co	0.614	1768	16,190	1675	1860	0.804
Cr	0.356	2180	16,900	1688	1860	0.446
Mo	0.030	2896	32,000	1765	1860	0.082

To determine the relative volatility (α_i) of the metal components present in the alloy, first the vapor pressure values for pure metals present in the alloy at T = 1860 K were calculated. For this purpose, regression equations were worked out (Table 3) from the graphic dependencies shown in Figure 2. The high correlation factor (R^2) shows that they can be successfully used to calculate the vapor pressure of pure metals in the temperature range of 1600 K to 2000 K. The integral value of the vapor pressure of the alloy at an operating temperature of 1860 K is also shown in the same table.

Table 3. Regression equations and vapor pressure of the metal elements and alloy at an operating temperature of 1860 K.

No	Equation	R^2	$P^o_{(Me)}$	$p_{j(Me)}$
1	$p_{Mn} = 2E - 6e^{0.009T}$	0.9951	3.72×10^1	-
2	$p_{Fe} = 1E - 13e^{0.0145T}$	0.9941	5.15×10^{-2}	-
3	$p_{Si} = 1E - 15e^{0.0154T}$	0.9935	2.74×10^{-1}	-
4	$p_{Nb} = 4E - 32e^{0.0271T}$	0.9958	3.10×10^{-10}	-
5	$p_W = 3E - 40e^{0.0324T}$	0.9959	4.433×10^{-14}	-
	$p_{Co} = 2E - 14e^{0.015T}$	0.9943	2.61×10^{-2}	2.1×10^{-2}
	$p_{Cr} = 6E - 13e^{0.0142T}$	0.9954	1.77×10^{-2}	7.89×10^{-3}
	$p_{Mo} = 3E - 28e^{0.0243T}$	0.9958	7.28×10^{-9}	5.97×10^{-10}
6	$P_{alloy} = p_{i(Co)} + p_{i(Cr)} + p_{i(Mo)} = 2.89 \times 10^{-2}$			

The values of the relative volatility α_i of the metal elements in relation to the CoCrMo alloy at T = 1860 K were calculated using Equation (1) and are shown in Figure 3.

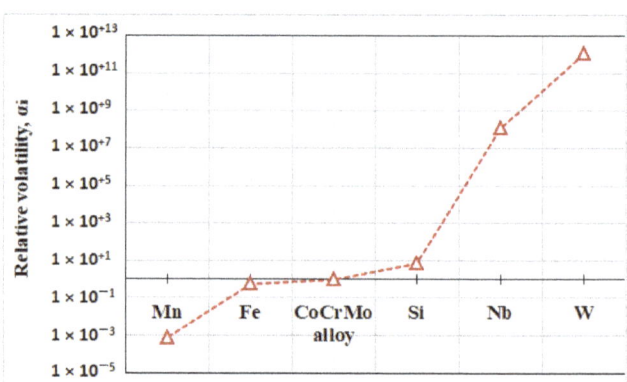

Figure 3. Values of the relative volatility α_i for the metal elements in CoCrMo alloy at 1860 K.

It can be seen that the α_i parameter varies in a wide range from 10^{-3} to 10^{12}. For more volatile elements such as Mn and Fe located to the left of the CoCrMo alloy, the values of $\alpha_i < 1$ and therefore their removal from the alloy is thermodynamically probable. For non-volatile elements (Nb and W) located on the right side of CoCrMo $\alpha_i \gg 1$ and therefore their removal is impossible. The removal of Si, whose value is $\alpha_{Si} \sim 1$, is possible, but it will be accompanied by significant losses of the alloy mass.

3.2. Refining Efficiency during EBMR of Technogenic CoCrMo Alloy

Table 4 shows the chemical composition of ingots after melting in different technological modes (heating temperature and residence time).

Table 4. Process parameters and chemical compositions (mass%) of the specimens before and after electron beam melting and refining.

Sample	Parameter		Concentration of Basic Elements				Concentration of Other Elements				
	T, K	τ, min	Co	Cr	Mo	Fe	Mn	Nb	W	Si	Others
Co-0	Initial alloy		61	31.22	4.78	0.65	0.43	0.32	0.38	1.09	0.13
Co-07	1790	20	62.11	31.01	4.96	0.37	0.0	0.31	0.37	0.73	0.14
Co-04	1830	10	62.60	30.51	4.91	0.35	0.0	0.31	0.36	0.83	0.13
Co-05	1830	20	62.92	30.41	4.90	0.31	0.0	0.31	0.36	0.67	0.12
Co-08	1830	30	63.54	29.79	4.99	0.27	0.0	0.31	0.36	0.62	0.12
Co-02	1845	20	63.99	29.35	5.05	0.29	0.0	0.31	0.36	0.56	0.09
Co-03	1860	10	64.14	29.81	4.80	0.18	0.0	0.31	0.35	0.37	0.04
Co-06	1860	20	64.94	28.79	5.06	0.12	0.0	0.31	0.35	0.38	0.05

The analysis of the results shows that as the temperature and retention time increase, the alloy is enriched in cobalt from 61% to 64.94%. The content of chromium decreases from 31.22% to 28.79%, and that of molybdenum slightly increases—from 4.78% to 5.06%. This can be explained keeping in mind that the vapor pressure of chromium is higher than that of cobalt, while the vapor pressure of molybdenum is much lower than that of Co and Cr (Figure 2).

For an easier interpretation of the behaviour of elements such as Fe, Mn, Si, Nb, and W, the influence of the temperature (Figure 4) and the retention time of 1830 K and 1860 K (Figure 5) on the degree of removal from CoCrMo alloy are expressed graphically. Mn is not shown in the figures as it is completely removed at 1790 K.

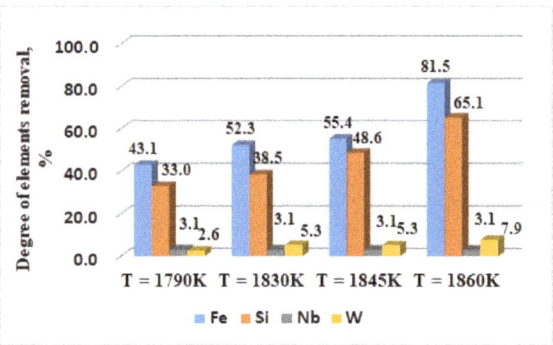

Figure 4. Influence of the melting temperature on the degree of alloying elements removal from the technogenic CoCrMo alloy.

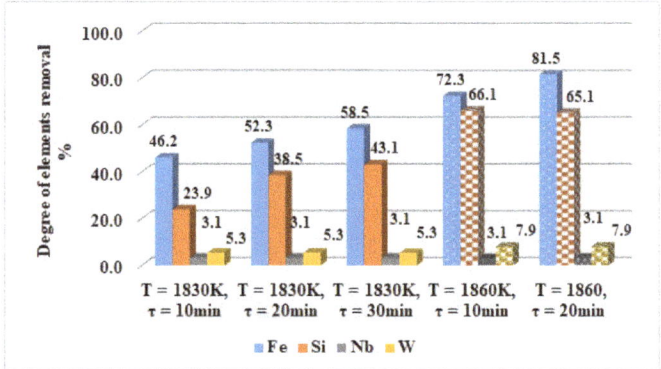

Figure 5. Influence of the melting time on the degree of elements removal at T = 1830 K and T = 1860 K from the technogenic CoCrMo alloy.

The influence of the temperature in the range from 1790 K to 1860 K at a retention time τ = 20 min on the degree of removal of the elements Fe, Si, Nb, and W is presented in Figure 4. The degree of refining (βi) is calculated using the equation:

$$\beta_{(i)} = \frac{C_{i(initial)} - C_{i(final)}}{C_{i(initial)}} \cdot 100\ \% \tag{5}$$

where $C_{i(initial)}$ and $C_{i(final)}$ are the initial and final content of the i-th element in the alloy, respectively.

The analysis of the results obtained indicates that the Fe and Si removal degree is constantly increasing with the increase in the temperature and at a maximum operating temperature (T = 1860 K, τ = 20 min) reaches 81.5% and 65.1%, respectively (Figure 4). With an extension of the retention time over 20 min, especially at a lower temperature (1830 K), the removal degree of the two elements changes negligibly. At a higher temperature and an extension of the retention time from 10 min to 20 min, the removal rate of iron (β_{Fe}) is increased by about 9% and hardly influences that of silicon (Figure 5). With regard to the more non-volatile elements such as Nb and W, the degree of removal is average ~3% for niobium and ~6% for W. Therefore, these elements cannot be refined by evaporation even at the highest temperature and extension of the melting time.

The conclusions drawn on the basis of the experiments carried out fully confirm the conclusions made on the basis of the thermodynamic analysis of the behaviour of the alloying elements present in the CoCrMo alloy.

The results obtained show that the highest degree of refining is achieved at T = 1860 K and a retention time of 20 min. The chemical composition of the alloy complies with the standard ASTM F75-12 for biomedical applications with the exception of tungsten, the concentration of which is slightly higher than the permissible amount. There are no requirements for the niobium content in the standard. Both alloying elements (W and Nb) improve the composition and structure of the alloy [3]. Tungsten stabilizes the hexagonal close-packed (HCP) structure and Nb stabilizes the face-centered cubic (FCC) structure [1].

3.3. Microstructures of CoCrMo Alloy after EBM

The main components in the studied CoCrMo alloy are Co (~61%), chromium (~31%), and molybdenum (~5%) and the phase changes that take place in the alloy during cooling after the EBM may be described with the Co–Cr–Mo phase diagram or the Co–Cr, Co–Mo, and Cr–Mo two-component diagrams.

Figure 6 shows the Co–Cr phase equilibrium diagram. At a Cr content of ~31%, the solidification of the alloy starts approximately at a temperature of 1670 K, avoiding the eutectic transition [31–33]. The high temperature γ-phase with a face-centered cubic (FCC) grid, which is stable up to ~1223 K, solidifies directly from the liquid. After that the martensitic transformation of the γ-phase starts up to a low temperature hexagonal close-packed (HCP) ε-phase. There is a eutectoid decomposition of the ε-phase → Co_3Cr + Co_2Cr at lower temperatures. At a higher concentration of Cr, the decomposition takes place along the reaction: ε-phase → Co_2Cr + Co_3Cr_2.

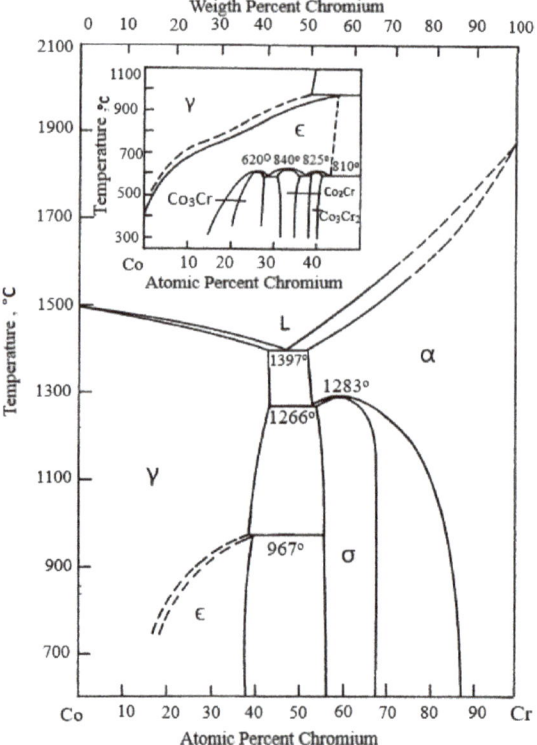

Figure 6. Two-component phase equilibrium diagram Co–Cr [31–33].

The Co–Mo diagram analysis shows that the Mo melting temperature is much higher than that of Co ($T_{m,Mo}$ = 2896 K, $T_{m,Co}$ = 1768 K) and the eutectic mixture was obtained at a temperature of 1608 K and approximately 40 mass% of Mo.

When examining Cr–Mo alloys it was found that chromium and molybdenum formed a spinoidal mixture. In the presence of Si, the molybdenum forms intermetallic compounds of the type of Mo_5Si_3 and Mo_3Si. At lower temperatures and a higher chromium content, σ-phase formation is observed. This phase is an intermetallic Co compound with Cr with a composition corresponding approximately to Co_2Cr_3 and in the presence of molybdenum—$Co_xCr_yMo_z$.

The impact of other alloying elements on the transformation temperature from the HCP to the FCC phase is summarized in [34] and elements such as Fe, Mn, Ni, Nb, and C reduce the temperature of the transformation from HCP to FCC, i.e., they are stabilisers of the FCC phase. Metals such as Cr, Mo, W, and Si increase the temperature of the transformation from HCP to FCC, therefore they are stabilisers for the HCP phase. These transformations are closely related to the alloy microstructure and hence, its mechanical and chemical properties.

Figure 7 shows a microstructure of samples of CoCrMo alloy before and after EBM at temperatures of 1790 K, 1845 K and 1860 K and a retention time of 20 min.

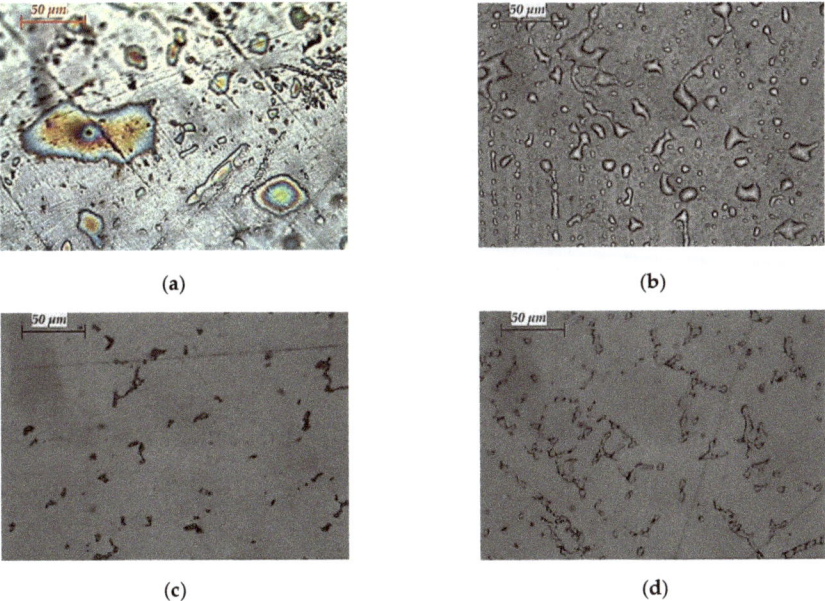

Figure 7. Optical micrographs of: (**a**) initial CoCrMo alloy; and ingots obtained after EBM: (**b**) T = 1790 K; (**c**) T = 1845 K; and (**d**) T = 1860 K; τ = 20 min (400× magnification).

The analysis of the microstructure of the starting sample indicates that it is highly oxidised and with a lot of defects. After an electron beam melting of the alloy at a temperature of 1790 K, a large number of intermetallic compounds situated on a Co matrix are observed on the surface. This can be explained by keeping in mind the low removal degree of the elements at this temperature (Figure 4).

With an increase in the temperature to 1860 K, the formation of a dendritic γ-phase, rich in cobalt, is observed. In this case, the degree of iron refining is very high (>80%), and that of Si is ~65%. W and Nb remain in the sample and it is only Nb that is a stabiliser of the γ-phase.

Microstructures of samples obtained at T = 1830 K and retention time τ = 10, 20 and 30 min are shown in Figure 8.

   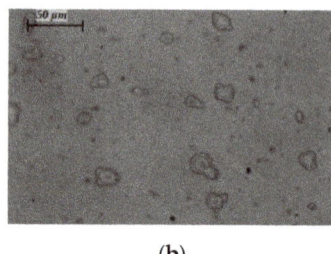   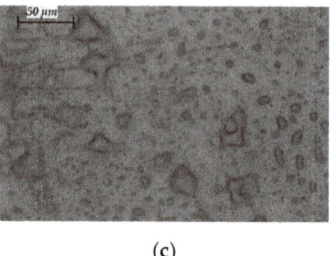

(a)   (b)   (c)

Figure 8. Microstructures of CoCrMo specimens manufactured at T = 1830 K for different refining time: (**a**) τ = 10 min; (**b**) τ = 20 min; and (**c**) τ = 30 min (400× magnification).

It is obvious that prolongation of the residence time results in the formation of a large number of intermetallic melts in the HCP phase. The transformation of the FCC structure to HCP is a very slow process and depends on both the melting parameters and the type of elements present in the alloy. The low refining degree of W and Si (Mn is completely removed at a temperature of 1790 K) from the CoCrMo alloy at T = 1830 K leads to the stabilisation of the low-temperature ε-phase [1]. This in turn leads to a decrease in the mechanical properties of the alloy and an increase in its corrosion resistance.

On the basis of the metallographic analysis using light microscopy, only a qualitative microstructure assessment of the samples can be provided. To obtain more accurate information on the influence of the EBM technological parameters on the microstructure of the ingots obtained, a microscopic study was performed using SEM and an elemental chemical analysis of the phases observed on the SEM/BEC images was made by EDS analysis. For each more characteristic phase (such as the colour and shape), at least 25 spectra are taken and the average chemical composition of the phase is calculated; they are labelled as "dark", "grey", and "light" phases, respectively.

Table 5 shows the chemical composition of the more characteristic phases observed on the SEM/BEC images of the primary CoCrMo alloy (Figure 9).

Table 5. EDS microanalysis of more characteristic phases observed in the initial CoCrMo alloy.

Sample	Area	Chemical Composition, at%							Phase
		Co	Cr	Mo	Fe	Si	Nb	W	
Co-0	A0 (dark)	61.57	29.99	7.19	0.42	0.83	0	0	$Co_{1.8}Cr(Mo, Fe, Si)$
	B0 (grey)	60.72	29.75	7.46	0.47	0.75	0.85	0	$Co_{1.8}Cr(Mo, Fe, Si, Nb)$
	C0 (light)	46.11	26.63	17.16	0	1.98	8.12	0	$Co_{0.8}Cr_{0.5}Mo_{0.2}(Si, Nb)$

The EDS analysis shows that the alloy's matrix is mainly based on the $Co_{1.8}Cr$ phase, close in composition to the Co_2Cr phase (Figure 6), which is obtained after the decomposition of an ε-phase at a temperature <973 K [31–33]. The differences between the "dark" and "grey" phases are mainly due to the different content of Mo (from 7.19 to 7.46 at%) and Nb (from 0 to 0.85 at%). The content of the other elements in these phases is <1 at%.

The "light" phase observed on the SEM photos is a molybdenum-rich phase ($Co_{0.8}Cr_{0.5}Mo_{0.2}$). The Nb and Si content in this phase is ~8 at% and ~2 at%, respectively.

Regardless of the large number of analyses, two elements were not identified—Mn and W. This can be explained by their low content in the studied alloy (Table 1).

Figures 10 and 11 show SEM/BEC images of the microstructure and the more characteristic intermetallic phases formed in the CoCrMo alloy after EBM at temperatures of

1830 K and 1860 K and retention time τ = 20 min. At a temperature of 1830 K, an experiment with a residence time of 30 min was performed (sample Co-08). The calculated values of the average chemical compositions of the phases observed on the SEM/BEC photographs are given in Table 6.

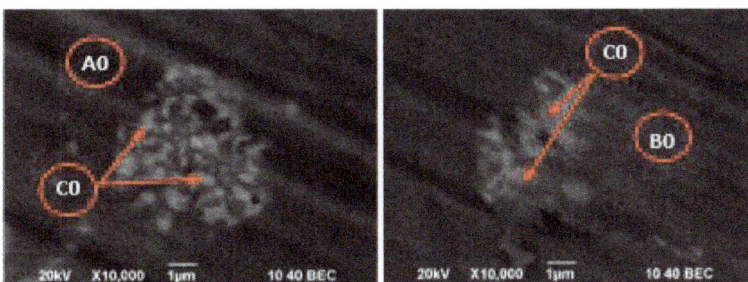

Figure 9. More characteristic phases observed on the SEM/BEC images of the CoCrMo material before e-beam melting (sample Co-0).

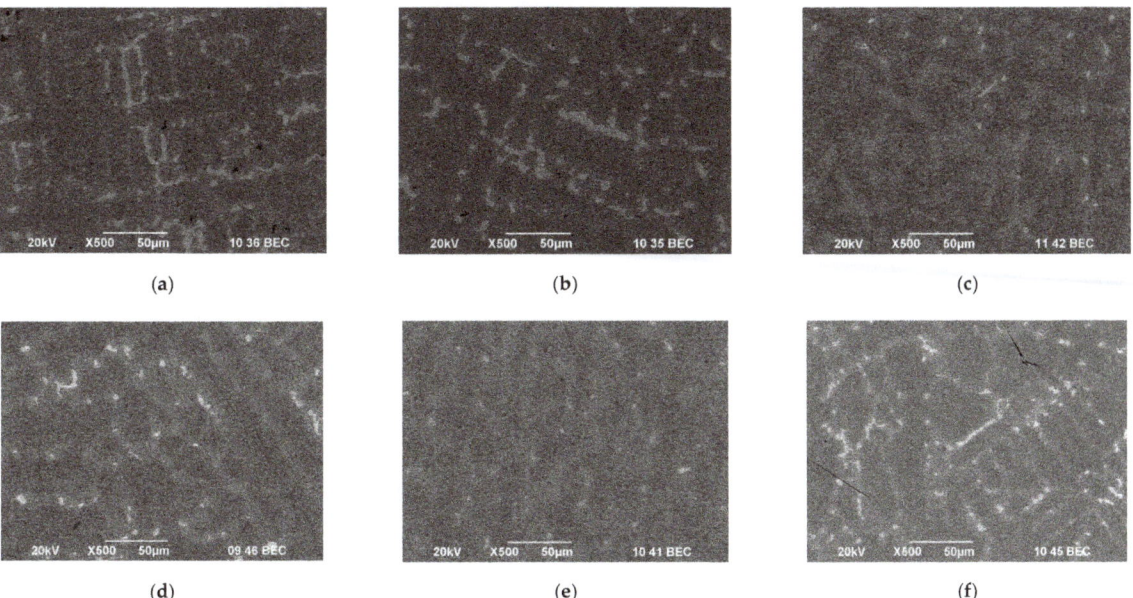

Figure 10. Microstructures of CoCrMo samples produced/manufactured at different process conditions: (**a**,**b**) T = 1830 K, τ = 20 min; (**c**,**d**) T = 1830 K, τ = 30 min; (**e**,**f**) T = 1860 K, τ = 20 min; (**a**,**c**,**e**)—microstructure on the periphery of the specimen; (**b**,**d**,**f**)—microstructure in the central part of the sample.

Figure 10 shows that the solidification process of CoCrMo alloy is accompanied by the formation of a dendritic structure and the formation of characteristic intermetallic melts. As the temperature increases, the solidification rate decreases, leading to the formation of the HCP structure. This process is more pronounced in the centre of the samples and shows that the process proceeds at the solid/liquid interface in the direction of direct heat removal.

The EDS microanalysis of the periphery and the centre of the samples (Table 6) shows non-essential differences in their chemical composition and therefore only the phases identified in the centre of the samples will be discussed. The influence of the temperature and the retention time on the chemical composition of the phases observed on the SEM images is represented in Figure 12.

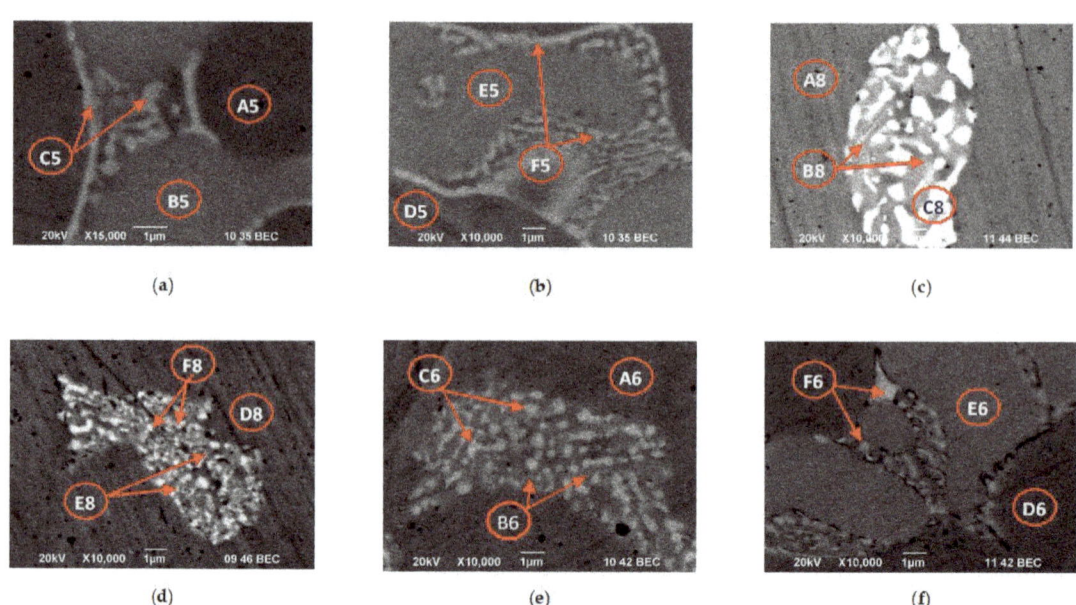

Figure 11. More characteristic phases observed in the CoCrMo specimens after EBM at different modes: (**a,b**) T = 1830 K, τ = 20 min; (**c,d**) T = 1830 K, τ = 30 min; (**e,f**) T = 1860 K, τ = 20 min; (**a,c,e**)—phases formed on the periphery of the sample; (**b,d,f**)—phases formed in the central part of the sample.

Table 6. Microanalysis of more characteristic phases observed at the periphery and in the center of CoCrMo ingots obtained after e-beam melting.

Sample		Area	Chemical Composition, at%						
			Co	Cr	Mo	Fe	Si	Nb	W
Co-05 T = 1830 K τ = 20 min	Rim	A5 (dark) B5 (grey) C5 (light)	59.73 48.30 53.45	33.55 36.83 28.67	4.87 11.06 9.78	0.23 0 0.11	1.62 2.06 3.25	0 1.75 4.74	0 0 0
	Core	D5 (dark) E5 (grey) F5 (light)	60.13 49.31 54.55	32.69 35.75 27.62	5.68 10.96 9.58	0 0 0	1.50 1.84 3.35	0 2.14 4.90	0 0 0
Co-08 T = 1830 K τ = 30 min	Rim	A8 (dark) B8 (grey) C8 (light)	61.24 50.30 50.65	32.41 31.81 32.07	4.66 11.03 11.23	0.17 0.52 0	1.52 2.85 2.60	0 3.49 3.45	0 0 0
	Core	D8 (dark) E8 (grey) F8 (light)	61.73 58.48 49.25	29.15 27.71 31.38	8.29 10.53 16.28	0 0 0	0.74 0.82 0.81	0.09 2.46 2.28	0 0 0
Co-06 T = 1860 K τ = 20 min	Rim	A6 (dark) B6 (grey) C6 (light)	61.19 59.35 54.23	29.56 29.02 23.85	8.28 8.89 13.68	0.38 0.45 0	0.59 0.72 1.59	0 1.57 6.65	0 0 0
	Core	D6 (dark) E6 (grey) F6 (light)	60.41 58.39 50.79	29.55 29.11 27.25	8.69 9.68 15.38	0.46 0.58 0.41	0.79 0.87 1.15	0.10 1.37 5.02	0 0 0

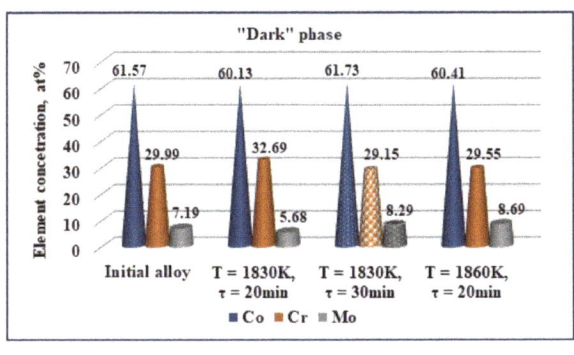

(a)

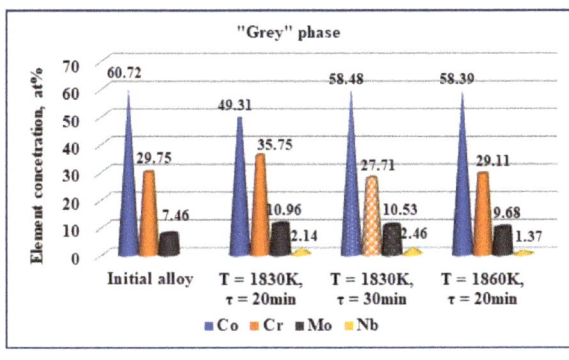

(b)

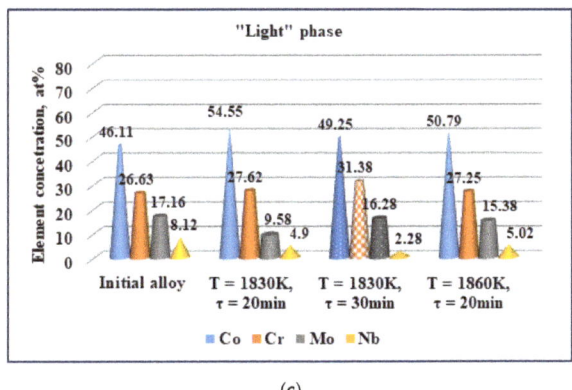

(c)

Figure 12. Influence of the temperature and melting time on the composition of the phases observed on the SEM pictures: (**a**) "dark" phase; (**b**) "grey" phase; and (**c**) "light" phase.

The EDS analysis shows a distinct micro-segregation in the cobalt matrix composition. At a temperature higher than 1830 K in the inter-dendrite zone ("dark"), a $Co_{1.8}Cr(Mo,Si)$ phase is formed. It is close in composition to the Co_2Cr phase, which is obtained after the decomposition of the ε-phase (Figure 6). The Co content of this phase is ~60 at% and does not change with an increase in temperature, while the Cr content decreases from ~33 at% to approximately 29 at% (Figure 12a). By increasing the temperature, the melt is enriched in Mo by ~3 at%. W is not present in this phase and the Nb content is less than 0.1 at%.

The chemical microanalysis of the "grey" phase shows a lower Co content and higher Mo content compared to the "dark" phase (Figure 12b). Unlike the "dark" phase, the Nb content is ~2 at%.

Out of the other elements present in the cobalt matrix, with the temperature increase, the Si content decreases both in the "dark" phase (from 1.5 to 0.79 at%) and in the "grey" phase (from 1.84 to 0.87 at%).

Chromium, molybdenum, and silicon diffuse to the inter-dendrite zones [35]. This in turn leads to an increase in the reflex's intensity. In our study it was observed that Nb also diffuses to the inter-dendrite zone of the material.

The chemical composition of the "light" phases shows that they are intermetallic compounds of the type $Co_xCr_yMo_z(Si,Nb)$. By increasing the temperature, the melt is enriched in Mo (from 9.58 to 15.38 at%) and Nb, while the Si content decreases (from 3.35 to 1.15 at%). In the studied temperature interval, the Cr content hardly changes, and it is ~27 at%, while the Co decreases slightly from 54 to 51 at%.

The EDS analysis of the "dark" phase shows that an extension of the retention time from 20 to 30 min (samples Co-05 and Co-08) results in a decrease in the Cr content by 3.5 at%, enriching in Mo from ~5.7 to ~8.3 at%. The change in Co content is negligible.

Unlike the "dark" phase, the Co content in the "grey" phase is significantly increasing (from ~49 to ~58.5 at%) with the extension of time, while that of Cr decreases from 35.7 to 27.7 at%. The content of Mo and Nb remains almost constant, and it is ~10 at% (Mo) and ~2 at% (Nb), respectively.

By extending the retention time, the chemical composition of the "light" phase is enriched in Cr (from ~27.6 to ~31 at%) and Mo (from 9.6 to 16 at%). The Nb and Si content in these melts decreases from ~4.9 to ~2.3 at% (Nb) and from ~3.3 to ~0.8 at% (Si), respectively.

4. Conclusions

This paper studies the possibility of refining a technogenic CoCrMo alloy using EBM. Based on the thermodynamic analysis, the vapor pressure of the major components of the alloy (Co, Cr, and Mo) and the other elements (Fe, Mn, Si, W, and Nb) present in it was calculated. The integral value of the vapor pressure of the alloy, taking into account the activities of the main alloy components and the relative volatility of the metal elements in the alloy were assessed. An experimental study of the influence of kinetic parameters (temperature and refining time) on the change in the chemical composition of the alloy and the microstructure of the ingots was carried out. The results obtained can be summarised as follows:

- Within the studied temperature range (1600–2000 K), it is most likely to have evaporation of Mn and Fe. Tungsten and niobium have significantly lower vapor pressures than Co, Cr, and Mo and cannot be removed from the alloy. Si has a vapor pressure close to that of Co and its evaporation is thermodynamically probable, but it will be difficult. Out of the main components of the alloy, the probability of evaporation of Cr is the highest.
- It has been found that the removal rate of Fe and Si increases with the increase in the temperature and with an extension of the retention time. It is shown that Mn is completely removed at the lowest operating temperature tested (1790 K) and the degree of removal of non-volatile elements (Nb and W) remains low. The highest degree of refining was achieved at 1860 K and a residence time of 20 min.
- The EDS analysis of the more characteristic phases observed on the SEM images of the samples shows a distinct micro-segregation in the matrix composition. At a temperature higher than 1830 K in the inter-dendrite "dark" zone, a $Co_{1.8}Cr(Mo,Si)$ phase is formed.
- It has been found that the "light" phases are intermetallic compounds of the type $Co_xCr_yMo_z(Si,Nb)$, which are enriched in Mo and Nb as the temperature increases.

The study has shown the possibility of recycling through EBM of CoCrMo technogenic material for the needs of biomedical practice.

Author Contributions: K.V. and V.S. contributed to the design of the study and interpretation of data; K.V. and V.V. conceived and designed the experiments; V.V. performed the experiments; S.A.-V. performed SEM images and EDS analysis; K.V., V.S. analyzed the data; K.V. and V.S. wrote the manuscript. All authors have read and agreed to the published version of the manuscript.

Funding: The work was supported by the Bulgarian National Science Fund under contract DN17/9.

Institutional Review Board Statement: Not applicable.

Informed Consent Statement: Not applicable.

Acknowledgments: The authors are grateful to M. Naplatanova and R. Nikolov for technical assistance in processing the samples.

Conflicts of Interest: The authors declare no conflict of interest.

References

1. Narushima, T.; Ueda, K.; Alfirano, A. *Advances in Metallic Biomaterials*; Niinomi, M., Narushima, T., Nakai, M., Eds.; Springer Series in Biomaterials Science and Engineering 3; Springer: Berlin/Heidelberg, Germany, 2015; pp. 157–178. [CrossRef]
2. Davis, J.R. (Ed.) *ASM Specialty Handbook: Nickel, Cobalt, and Their Alloys*; Materials Park; ASM International: Almere, The Netherlands, 2000; p. 356.
3. Mendes, P.S.N.; Lins, J.F.C.; Mendes, P.S.N.; Prudente, W.R.; Siqueira, R.P.; Pereira, R.E.; Rocha, S.M.S.; Leoni, A.R. Microstructural characterization of Co-Cr-Mo-W alloy as casting for odontological application. *Int. J. Eng. Res. Appl.* **2017**, *7 Pt 1*, 34–37. [CrossRef]
4. Ghiban, A.; Moldovan, P. Study of corrosion behavior under simulated physiological conditions of the dental CoCrMoTi alloys. *UPB Sci. Bull. Ser. B* **2012**, *74*, 203–214.
5. Dobruchowska, E.; Paziewska, M.; Przybyl, K.; Reszka, K. Structure and corrosion resistance of Co-Cr-Mo alloy used in Birmingham Hip Resurfacing system. *Acta Bioeng. Biomech.* **2017**, *19*, 31–39. [PubMed]
6. Matkovic, T.; Matkovic, P.; Malina, J. Effects of Ni and Mo on the microstructure and some other properties of Co-Cr dental alloys. *J. Alloys Compd.* **2004**, *366*, 293–297. [CrossRef]
7. Yıldırım, M.; Keleş, A. Production of Co-Cr-Mo biomedical alloys via investment casting technique. *Turk. J. Electromech. Energy* **2018**, *3*, 12–16.
8. Gulisija, Z.; Patarıc, A.; Mihailovic, M. Co-Mo-Cr alloys for dentistry obtained by vacuum precise casting. *Zastita Materijala* **2015**, *56*, 175–178. [CrossRef]
9. Lee, S.-H.; Takahashi, E.; Nomura, N.; Chiba, A. Effect of heat treatment on microstructure and mechanical properties of Ni- and C-free Co–Cr–Mo alloys for medical applications. *Mater. Trans.* **2005**, *46*, 1790–1793. [CrossRef]
10. Wei, D.; Koizumi, Y.; Takashima, T.; Nagasako, M.; Chiba, A. Fatigue improvement of electron beam melting-fabricated biomedical Co–Cr–Mo alloy by accessible heat treatment. *Mater. Res. Lett.* **2018**, *6*, 93–99. [CrossRef]
11. Wei, D.; Anniyaer, A.; Koizumi, Y.; Aoyagi, K.; Nagasako, M.; Kato, H.; Chiba, A. On microstructural homogenization and mechanical properties optimization of biomedical Co-Cr-Mo alloy additively manufactured by using electron beam melting. *Addit. Manuf.* **2019**, *28*, 215–227. [CrossRef]
12. Porojan, S.; Birdeanu, M.; Savencu, C.; Porojan, L. Structural and morphological approach of Co-Cr dental alloys processed by alternative manufacturing technologies. *J. Phys. Conf. Ser.* **2017**, *885*, 012005. [CrossRef]
13. Kim, H.R.; Jang, S.-H.; Kim, Y.K.; Son, J.S.; Min, B.K.; Kim, K.-H.; Kwon, T.-Y. Microstructures and mechanical properties of Co-Cr dental alloys fabricated by three CAD/CAM-based processing techniques. *Materials* **2016**, *9*, 596. [CrossRef] [PubMed]
14. Kircher, R.S.; Christensen, A.M.; Wurth, K.W. Electron beam melted (EBM) Co-Cr-Mo Alloy for orthopaedic implant applications. In Proceedings of the International Solid Freeform Fabrication Symposium, Austin, TX, USA, 3–5 August 2009; pp. 428–436.
15. Guoqing, Z.; Junxin, L.; Xiaoyu, Z.; Jin, L.; Anmin, W. Effect of heat treatment on the properties of CoCrMo alloy manufactured by selective laser melting. *J. Mater. Eng. Perform.* **2018**, *27*, 2281–2287. [CrossRef]
16. Sun, S.-H.; Koizumi, Y.; Kurosu, S.; Li, Y.-P.; Chiba, A. Phase and grain size inhomogeneity and their influences on creep behavior of Co–Cr–Mo alloy additive manufactured by electron beam melting. *Acta Mater.* **2015**, *86*, 305–318. [CrossRef]
17. Tan, X.P.; Wang, P.; Kok, Y.; Toh, W.Q.; Sun, Z.; Nai, S.M.L.; Descoins, M.; Mangelinck, D.; Liu, E.; Tor, S.B. Carbide precipitation characteristics in additive manufacturing of Co-Cr-Mo alloy via selective electron beam melting. *Scr. Mater.* **2018**, *143*, 117–121. [CrossRef]
18. Mladenov, G.; Koleva, E.; Vutova, K.; Vassileva, V. *Practical Aspects and Application of Electron Beam Irradiation*; Nemtanu, M., Brasoveanu, M., Eds.; Transworld Research Network: Trivandrum, India, 2011; pp. 43–93.
19. Kalugin, A. *Electron Beam Melting of Metals*; Metallurgy Publishing House: Moscow, Russia, 1980. (In Russian)
20. Ladokhin, S.; Levitcky, N.; Lapshuk, T.; Drozd, E.; Matvietc, E.; Voron, M. The electron-beam melting use for medical cast parts production. *Met. Cast. Ukr.* **2015**, *263*, 7–11. (In Russian)

21. Vutova, K.; Vassileva, V.; Stefanova, V.; Naplatanova, M. Influence of process parameters on the metal quality at electron beam melting of molybdenum. In *11th International Symposium on High-Temperature Metallurgical Processing*; Peng, Z., Hwang, J., Downey, J.P., Gregurek, D., Zhao, B., Yücel, O., Keskinkilic, E., Jiang, T., White, J.F., Mahmoud, M.M., Eds.; The Minerals, Metals & Materials Series; Springer: Cham, Germany, 2020; pp. 941–951.
22. Vutova, K.; Stefanova, V.; Vassileva, V.; Kadiyski, M. Behaviour of impurities during electron beam melting of copper technogenic material. *Materials* **2022**, *15*, 936. [CrossRef]
23. Vutova, K.; Vassileva, V.; Stefanova, V.; Amalnerkar, D.; Tanaka, T. Effect of electron beam method on processing of titanium technogenic material. *Metals* **2019**, *9*, 683. [CrossRef]
24. Vutova, K.; Vassileva, V.; Koleva, E.; Georgieva, E.; Mladenov, G.; Mollov, D.; Kardjiev, M. Investigation of electron beam melting and refining of titanium and tantalum scrap. *J. Mater. Process. Technol.* **2010**, *210*, 1089–1094. [CrossRef]
25. Vander Voort, G.F. Metallorgraphy of Superalloys. *Ind. Heat.* **2003**, *70*, 40–43.
26. HSC Chemistry, v. *HSC Chemistry v. 7.1*; Metso Outotec Research Center: Pori/Helsinki, Finland, 2013.
27. Bobrov, U.P.; Virich, V.D.; Dimitrenko, A.E.; Koblik, D.V.; Kovtun, G.P.; Manjos, V.V.; Pilipenko, N.N.; Tancura, I.G.; Shterban, A.P. Refining of ruthenium by electron beam melting. *Quest. At. Sci. Technol. Vac. Pure Mater. Supercond.* **2011**, *6*, 11–17. (In Russian)
28. Kubaschewski, O.; Alcock, C.B. *Metallurgical Thermochemistry*, 5th ed.; Pergamon Press Ltd.: Oxford, UK, 1979.
29. Yao, K.; Min, M.; Shi, S.; Tan, Y. Volatilazation behaviour of β-type Ti-Mo alloy manufactured by electron beam melting. *Metals* **2018**, *8*, 206. [CrossRef]
30. Manasijevic, D.; Zivkovic, D.; Zivkovic, Z. Calculation of activities in Ga-Cd and Cu-Pb binary systems. *J. Min. Metall.* **2002**, *38*, 273–284. [CrossRef]
31. Allibert, C.; Bernard, C.; Valignat, N.; Dombre, M. Co-Cr binary system: Experimental Redetermination of the Phase Diagram and Comparison with the Diagram Calculated from the Thermodynamic Data. *J. Less Common Met.* **1978**, *59*, 211–228. [CrossRef]
32. Gupta, K.P. *Phase Diagrams of Ternary Nickel Alloys Part I*; The Indian Institute of Metals: Kolkata, India, 1990; p. 349.
33. Gupta, K.P. The Co-Cr-Mo (cobalt-chromium-molybdenum) system. *J. Phase Equilibria Diffus.* **2005**, *26*, 87–92. [CrossRef]
34. Beltran, A.M. *Superalloys II*; Sims, C.T., Stoloff, N.S., Hagel, W.C., Eds.; Wiley: New York, NY, USA, 1987; pp. 135–163.
35. Podrez-Radziszewska, M.; Haimann, K.; Dudzinski, W.; Morawska-Soltysik, M. Characteristic of intermetallic phases in cast dental CoCrMo alloy. *Arch. Foundry Eng.* **2010**, *10*, 51–56.

MDPI
St. Alban-Anlage 66
4052 Basel
Switzerland
Tel. +41 61 683 77 34
Fax +41 61 302 89 18
www.mdpi.com

Materials Editorial Office
E-mail: materials@mdpi.com
www.mdpi.com/journal/materials

www.ingramcontent.com/pod-product-compliance
Lightning Source LLC
LaVergne TN
LVHW070628100526
838202LV00012B/752